CAMBRIDGE COUNTY GEOGRAPHIES
SCOTLAND
GENERAL EDITOR: W. MURISON, M.A.

BANFFSHIRE

Cambridge County Geographies

BANFFSHIRE

BY

W. BARCLAY

Editor, *The Banffshire Journal*

With Maps, Diagrams and Illustrations

CAMBRIDGE
AT THE UNIVERSITY PRESS
1922

CAMBRIDGE UNIVERSITY PRESS
Cambridge, New York, Melbourne, Madrid, Cape Town,
Singapore, São Paulo, Delhi, Mexico City

Cambridge University Press
The Edinburgh Building, Cambridge CB2 8RU, UK

Published in the United States of America by Cambridge University Press, New York

www.cambridge.org
Information on this title: www.cambridge.org/9781107614949

First published 1922
First paperback edition 2013

A catalogue record for this publication is available from the British Library

ISBN 978-1-107-61494-9 Paperback

CONTENTS

ILLUSTRATIONS

ILLUSTRATIONS

MAPS

ILLUSTRATIONS

The illustrations on pp. 4, 16, 35, 96, 125 are reproduced from photographs by Mr W. Gammie; those on pp. 7, 11, 12, 129 from photographs by Mr G. MacLennan; those on pp. 13, 29, 105, 134 from photographs by Mr J. R. Gordon; those on pp. 18, 20, 92, 98 from photographs by Messrs J. Valentine & Sons Ltd.; that on p. 21 from a photograph by Mr G. Laing; that on p. 24 is reproduced by permission of H. M. Geological Survey, Scotland; those on pp. 37, 39, 41, 43, 90, 108 are from photographs by Mr T. Newton; that on p. 56 from a photograph supplied by the Aberdeen Angus Soc.; those on pp. 62, 67, 69, 102, 132 from photographs by Mr H. Holman; those on pp. 64, 86 from photographs by Mr P. T. Clark; those on pp. 75, 83 [2], 126 from photographs supplied by *The Banffshire Journal*; those on pp. 82, 83, 84, 91 are reproduced by permission of The Society of Antiquaries of Scotland; those on pp. 89, 97, 130 are from photographs by Mr R. B. Newton; that on p. 94 from a photograph by the Rev. R. H. Calder; that on p. 100 was supplied by the Rev. H. D. F. Dunnett; that on p. 101 is from a photograph by Messrs G. Pirrie & Sons; that on p. 104 from a photograph by Mr W. Pearson; that on p. 113 is reproduced by permission of the Clarendon Press, Oxford; that on p. 122 by arrangement with Messrs Cassell & Co. Ltd., and that on p. 127 is from a photograph by Mr J. P. Pozzi.

BANFFSHIRE

1. County and Shire. The Origin of Banff.

The word *shire* is of Old English origin and meant office, charge, administration. The Norman Conquest introduced the word *county*—through French from the Latin *comitatus*, which in mediaeval documents designates the shire. *County* is the district ruled by a count, the king's *comes*, the equivalent of the older English term *earl*. This system of local administration was in England the result of a gradual, orderly and natural development; in Scotland, on the other hand, it was the result of the administrative Act of David I (1124–53), who, by residence in England was so "polished from a boy" that "he had rubbed off all the rust of Scottish barbarity." With an intimate knowledge of English methods of administration he sought to introduce some of these. He accordingly divided Scotland into sheriffdoms. This step marked the beginning of the Scottish county division as it is known today, although it took a long time to complete, for the Celtic chiefs in the north and in Galloway were as yet too powerful to allow royal officials to hold courts within their territories. The policy of David, however, led to the all but complete expulsion of the Celtic system from the whole of the east of Scotland up to the Moray Firth, including a not inconsiderable portion of Banffshire.

Originally the civil counties were synonymous with the sheriffdoms or stewartries, the stewartry ceasing with the abolition of hereditary jurisdictions in 1748. By the Act of David, Scotland was divided into 25 sheriffdoms or counties. In the latter part of the thirteenth century they numbered 34; there are now 33.

The county of Banff existed at an early period of the new regime. In the twelfth century and in the thirteenth we find such varied forms of its name as *Banb*, *Banef*, *Bamphe*, *Banffe*, *Banet*. Curiously divergent derivations have been given. The Celtic words for "white ford or beach," for "sucking-pig," and for "holy woman," have been suggested. Banba, a Welsh or Irish queen, has also been mentioned as bestowing her name. Amid such divergencies, who shall decide?

2. General Characteristics, Position and Natural Conditions.

We often speak of an imaginary line from Helensburgh to Stonehaven as marking off the Highlands from the Central Lowlands. This, however, is not the whole truth. For, while part of Banffshire is certainly highland, its northern part is really lowland—a statement holding good also for its eastern and western neighbours. This lowland region on the Moray Firth is geologically, topographically and meteorologically different from the highland region to the south, and consequently differs considerably in density of population, in products and in the occupations of the inhabitants.

Banffshire lies between latitude 57° 6′ and 57° 42′ north, and between longitude 2° 15′ and 3° 40′ west. To the east and south it has Aberdeenshire; to the west the shires of Inverness and Moray.

Near the coast the surface is comparatively level and is mostly of fine, open, undulating country, of rich, highly cultivated soil. In the south and south-east it is mountainous, with extensive and good farms however in the fertile glens. The chief mountain ranges, rivers and strike of the stratified rocks, run from south-west to north-east, the whole county being an extensive slope in the same direction, from the Grampians to the Moray Firth.

In the south are productive deer-forests, and some of the grouse-moors, extending to tens of thousands of acres, are among the finest in Scotland.

In other aspects, too, Banffshire stands pre-eminent. Despite its comparatively short sea-board, it has a greater wealth than any other county in herring-fishing plant and stands supreme in the size and the value of its herring-fishing fleet, propelled either by steam or motor engine. Along its shores, from the bay of Gamrie westward to Portgordon, is the largest aggregation of herring and line fishermen, who, in the herring fisheries of Scotland, and only to a less extent in those of England and Ireland, exercise a decisive influence. Not a few fishermen from the county were selected by the Congested Districts Board to introduce and teach Scottish fishing methods in Ireland, and requests have come from Japan for them to undertake similar work there. The manufacture of malt whisky represents a large and important interest, and probably the county output of spirits is the largest in Scotland. For many years Banffshire

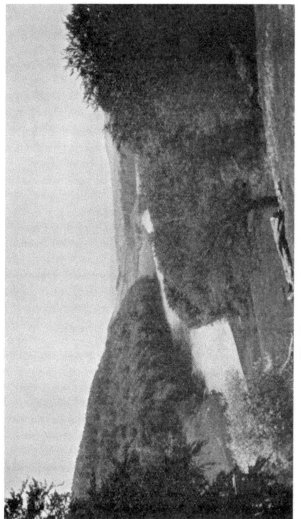

The Deveron at Netherdale, Marnoch

has possessed an advanced system of agriculture, and farming may be taken as the leading industry. With a comparatively mild climate along a considerable width of the seaboard, agricultural conditions are wonderfully favourable. But in districts such as Glenrinnes, the Cabrach, Kirkmichael and parts of Glenlivet, where cultivated land climbs from the valleys up the hillsides and abuts on the heather, the more or less absolute failure of a crop is not unknown, and in some years in these high altitudes the storms of the northern December see cereal crops lying unsecured, destroyed by weather and consumed by game. In the large parish of Cabrach, an extensive area of which is moorland, hill and forest, is the farm of Reekimlane: it was the only "reekin' lum" left in the parish in a time of physical stress and hardship caused by crop failures. Besides agriculture, fishing and distilling, there are minor industries —boat-building, tweed manufacture, the making of agricultural implements, lime-burning and the like; but these are not of the same general importance.

Banffshire has for long been noted for its love of education, and the most potent export indeed, is not its whisky, its black cattle, or its herrings, but young men and women fitted by education and discipline to play a creditable part in the affairs of life. The teachers of the county enjoy the benefit of the bequest of James Dick, a West Indian and London merchant, born at Forres in 1743, who died in 1828, and left £113,000 to promote higher learning among the parish schoolmasters of Aberdeenshire, Banffshire and Moray. The influence of the bequest has been most beneficial in encouraging country schools to maintain a high standard of education. Such schools as those of Banff,

Fordyce, Keith, Tomintoul and others have for many years occupied an important place in the higher educational activities of the county, and through them there is maintained an intimate connection with the University of Aberdeen. At a meeting in the county, Professor Laurie, after an experience of 35 years as Dick Bequest Visitor, said that he had some knowledge of what was going on in America, Germany and France and he would assure them it was a fact, as he had stated, that Banffshire stood quite at the head of all educational effort and machinery and efficiency of any part of the civilised world he knew of or read of.

The magnificent sea-cliffs and the fine sea-views attract artists. Many visitors come annually to the bathing-places and the golf links with their bracing air—bracing it must be for the breezes sweep off the sea straight from the Arctic Zone with no land between the Banffshire coast and the North Pole. Inland, too, the summer visitor resorts to places like Dufftown and Tomintoul, while in Cairngorm and Ben Macdhui the mountaineer finds fit kingdoms to conquer. The geologist and the naturalist will also discover much of interest in the county.

3. Size. Shape. Boundaries.

Banffshire, with an area of 403,053 acres, about 630 square miles, stands fourteenth among Scottish counties. From north-east to south-west, it is 67 miles long; its greatest width, which is along the coast, is 32 miles, but at Keith, near the centre of the county, it narrows to about

Dufftown

nine miles, again expanding southward, so that in shape
it may be said to resemble an hour-glass.

Previous to the Local Government Act of 1889, of
thirty civil parishes in Banffshire, eighteen were wholly

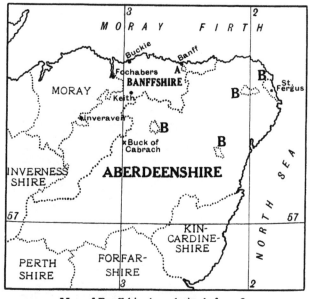

Map of Banffshire boundaries before 1890
(*Note detached portions* (B) *in Aberdeenshire, and one detached portion*
(A) *of Aberdeenshire in Banffshire.*)

within the county, portions of six were in Aberdeenshire,
and five in Moray, while one was wholly detached, the
parish of St Fergus in Eastern Buchan. It was originally
the property of a family who, as hereditary Sheriffs of Banff
were naturally desirous to have their domain within their

own jurisdiction and were able to secure its annexation to
Banffshire; but this feudal peculiarity ceased in 1890. At
the same time the other parishes belonging to different shires
were transferred to one. Thus the parishes of St Fergus,
Old Deer, New Machar, Gartly and Glass became wholly
Aberdeenshire, while Bellie and Rothes were placed alto-
gether in Moray. The whole parishes of Cabrach, Boharm,
Inveraven and Keith were transferred to Banff. The
Banffshire portions of Cairnie and King Edward were
attached to other parishes of Banffshire.

The Moray Firth forms the boundary from the Tynet
Burn to the Tore Burn, where the county marches with
Aberdeenshire. The boundary then runs south and west
in a sinuous line to the Deveron near Eden House. Onwards to where the Isla joins the Deveron, near Rothiemay,
the river sometimes is and sometimes is not the dividing
line. As far as Grange station the Isla is the boundary,
which then mounts the watershed between Deveron and
Spey, and, sweeping past Glass, crosses the Deveron and
continues by the Buck of the Cabrach round by Ben Aven
and Ben Macdhui to the skirts of Braeriach, where it bids
goodbye to Aberdeenshire. Meeting Inverness-shire and
turning to the north-east, the boundary passes Cairngorm
and twists north-east, north and north-west to the Crom-
dale Hills, where it touches Moray. It holds north to the
Spey near Ballindalloch Station. Except for a short distance
round Ben Aigan, the river is the march till near Fochabers.
Then the boundary goes in an irregular line by Thief's Hill,
and zigzags to the Tynet Burn, along which it runs to the
sea.

4. Surface Features.

In the northern part of Banffshire there are hills that serve as useful landmarks at sea, while the southern possesses some of the highest mountains in Great Britain. One characteristic height in the north is the Bin of Cullen (1050 feet), with its neighbours the Little Bin and the Hill of Maud. From the top of the conical Bin the spectator has a fine panorama of sea and land. To the south the prospect stretches to Cairngorm, to the west are the mountains of Inverness-shire, to the north the coast land lies at his feet with headlands and bays, villages and towns, while across the Firth his eye rests on the Sutors of Cromarty, Ben Wyvis and other hills of Northern Scotland. Even lower heights are interesting watch-towers. From the Hill of Alvah (578 feet) we may see a large tract of Buchan, its somewhat monotonous aspect relieved by the bold headlands of Gamrie and Troup; to the south thriving woods and fertile lands with Benachie in the distance; the Buck of the Cabrach and Ben Rinnes in the south-west; to the north the wooded park of Duff House and the town of Banff; and beyond the sea the fantastic forms of the Caithness Hills.

The Knock Hill (1409 feet) dominates a large area in the lower part of the county. Ben Aigan, looking down on Banff and Moray and swept by the Spey, rises to 1544 feet. Further up the valley, Ben Rinnes rears its head to a height of 2755 feet, with its less exalted neighbours, the Meikle and Little Convals, while the adjoining Forests of Glenfiddich and Blackwater have heights well

over 2000 feet. Eastward the Buck of the Cabrach
(2368 feet) stands sentinel, while to the west, in the direction of the Braes of Glenlivet, are the Ladder Hills
(2475 feet), over which runs the mountainous road to the
upper valley of the Don.

From this point southwards is an extensive spur of the

Ben Rinnes

Grampians, peak upon peak rising in view amid the waste
of mountains. Many of them are between 2000 and
3000 feet. West of Inchrory is Garravoun (2431 feet);
in the Forest of Glenaven is the Bruach (2338 feet);
between the Gairn and the Aven, Ben Aven towers to a
height of 3843 feet; the Cairngorm group on the confines of
the county with Inverness is itself dominated by Cairngorm

Meikle Conval, Dufftown

(4084 feet) and where the county meets Aberdeen is the mighty mass of Ben Macdhui (4296 feet). Among these hills are the infant waters of the Dee, the Don, the Aven and many smaller streams, some reaching the North Sea at Aberdeen; others flowing to the Moray Firth.

In these wild regions, winter tarries long. From Tomintoul one may see extensive patches of white at midsummer,

Inchrory Lodge, Kirkmichael

and autumn is not gone when the hills around have got the covering of a new winter's snow. It is the land of the ptarmigan, the white hare, the lordly buck, and the peregrine; and in its lower altitudes the rifle takes its toll of the fox in his predatory tours among the flocks of hardy black-faced sheep that here find their summer home. The eagle is not yet extinct in the immense Forest of Glen Aven. The "beat" of the single policeman at Tomintoul includes the Cairngorms, but probably he does not very frequently take so wide a circuit. The same uninhabited area appertains to the ecclesiastical parish of Tomintoul, which must surely be one of the most extensive *quoad sacra* parishes in Scotland, including as it does about nine miles in length of the inhabited part of the civil parish of Kirkmichael, and the twelve or fifteen uninhabited miles that stretch into the Grampians.

5. Rivers and Lochs.

While the county is well watered, it possesses no great river entirely its own flowing directly into the sea. The Deveron is shared with Aberdeen, the Spey with Inverness and Moray, though the main drainage area of the former is in Banffshire, and from Banffshire comes the largest tributary of the latter.

"I hae a kintra," runs a rhyme attributed to Jane Maxwell, Duchess of Gordon—

> I hae a kintra caa'd the Cabrach,
> The folks dabrach,
> The water's Rushter,
> An' the corn's trushter.

In its comparative isolation the Cabrach is a little territory by itself, hence the "the" used before the name; it is called never Cabrach but "the" Cabrach. It is in the wild recesses of the Cabrach that the infant Deveron originates, in a land of heath-covered hills, of barren moors, and of far-stretching, rugged deer forests. The climate is unkindly, winter lingers long, so that, while the valleys are devoted to a somewhat precarious course of arable farming, it is a district of cattle- and sheep-rearing rather than of grain-growing. From the Cabrach to Rothiemay the Deveron flows through Aberdeenshire. The stream in its upper stretches runs rapidly along a series of glens and is frequently subject to violent freshets. All the bridges above Huntly were swept off by the historic floods of 1829. At Huntly it is joined by the Bogie.

At the point where the Deveron again touches Banff-shire, it receives the waters of the Isla, which issues from beautiful Loch Park, and runs through the parishes of Botriphnie, Keith and Grange. The Deveron now goes eastward by Marnoch, passing finely situated mansion houses, on to Inverkeithney, where the Burn of Forgue enters, coming from storied Frendraught and Glendronach. Still holding east, it nearly reaches Turriff but turns very abruptly northward. At the elbow it is joined by the Water of Turriff. Near the same point it is spanned by a bridge of three arches, of Delgaty freestone. The river flows on past Forglen House, in charming scenery, Mountblairy and Denlugas, to receive the Burn of King Edward. This stream comes from the east along the valley of King Edward. One of its branches begins near the church of Gamrie, within a short distance of the sea, and after a course of nine miles joins the Deveron at a point five miles from its mouth.

The Deveron now passes the ruins of Eden Castle, and, turning westward, enters the picturesque and romantic narrows between the Hill of Alvah and the Hill of Mont-coffer. Here a precipitous chasm is spanned by a bridge erected by the Earl of Fife. The chasm under the bridge is narrowed by the rocks to 27 feet, while the depth of the

The Deveron at Drachlaw

water is 50 feet. To the north of the bridge the rocks recede, rising to 100 feet above the water, and are fringed and covered with a rich diversity of shrubs and trees. Soon a fine valley gradually opens out. The river sweeps round its eastern side and encloses the plain on which Duff House stands. Half a mile hence it reaches the sea beneath the "Bonnie Brig' o' Banff." Until 1763 the river was crossed by fords and ferry boats. The first bridge was destroyed by a flood in 1768. The present bridge, designed by Smeaton (of the Eddystone lighthouse), was opened in 1780, and was widened in 1881. It is a beautiful structure of seven arches, and has a free waterway of 142 yards. The length of the Deveron is just short of 62 miles.

The Spey is nearly 50 miles longer than the Deveron and drains an area of over 1200 sq. miles. It rises at a great height above sea-level and receives a huge volume of water from numerous tributaries; and thus in its lower reaches it is the swiftest of Scottish rivers.

The Spey touches Banffshire close to Ballindalloch Station and soon after receives, at Inveraven, its largest and most beautiful tributary—the Aven, locally the A'an. The Aven flows entirely through Banffshire territory, traversing in its course of about 40 miles some of the finest scenery in the county, almost matchless for wild and rugged grandeur. It is a deep and rapid stream, clear as crystal.

> The water o' A'an so fair and clear,
> Would deceive a man of a hundred year.

It has its source on Ben Macdhui and issues from Loch Aven, already a considerable stream, flows through the entire length of the parish of Kirkmichael and falls into the Spey in the adjoining parish of Inveraven. Near Delnabo

it is joined by the Water of Ailnach; and north of Tomintoul
by the Conglass, from the hills overlooking Strathdon. A
little further on it receives the water of the Chabet, and in
the kindlier region of Glenlivet the Livet, swollen here into
a considerable stream by the tributaries of Crombie and
Tervie. The former, which drains the Braes of Glenlivet,
falls into it at Tombae, and the latter, which drains the

Loch Aven and Ben Macdhui

district of Morinsh and the lands bordering on Glenrinnes,
at Tombreakachie. The Aven, at its beginning, is about
4000 feet above the sea; at the Builg Burn, where it turns
north, 1300 feet. At St Bridget, Tomintoul, it has fallen
to 1000, at the Livet to 600, while the haugh at its junction with the Spey is 500 feet above the sea. The greater
part of the Aven valley is interspersed with natural-growing
birch and alder, which adds much to the grace and beauty

of the scenery and makes a tour by the road skirting its banks an ever pleasant memory.

From Inveraven the Spey flows past Aberlour, and on through beautiful scenery to Craigellachie, where it is joined by the Fiddich. The Fiddich rises in Glen Fiddich, the fine valley where the Duke of Richmond and Gordon has his deer forest and hunting lodge. The Dullan descends from Glenrinnes, the valley to the west of Glen Fiddich. The two streams unite at Tininver ("the meeting of the waters") below the church of Mortlach, about five miles from the Spey. Their whole course is about twelve or fourteen miles.

From Craigellachie, the Spey, amid superb scenery, rushes in majestic curves past the gloriously wooded heights of Arndilly. At Ordiequish it leaves Banffshire and dashes through Moray to the sea.

For salmon the Spey stands high; for trout it falls below the Aven, which is in great repute with anglers. The Deveron is famous for both trout and salmon. As far as records go the heaviest Deveron salmon was landed from the Eden water on the last day of the season of 1920. It weighed 56 lbs., and measured 53 inches in length, and 29 inches in girth. In the same section of the river in the spring of 1920 a sea-trout of 16½ lbs. was landed, the heaviest Deveron trout recorded.

The lakes in the county are not numerous, but fewness is compensated for by picturesqueness. Loch Park, in Botriphnie, is among the most charming of Highland lochs. Its beauty is familiar to many, since the railway skirts its shores. About a mile long, with a mean breadth of 100 yards, Loch Park occupies the base of a mountain

Aberlour—from Wester Elchies

gorge, the wooded sides of which rise to a considerable height. Here is the watershed of the district. The Isla, issuing from the east end of the loch, runs to the Deveron, while streamlets to the west of this glen reach the sea through the channel of the Spey.

The Shelter Stone, Ben Macdhui

Loch Aven, three miles long and one mile broad, is in the southern extremity of the parish of Kirkmichael and lies picturesquely amid wild and magnificent scenery. The towering heights of Ben a Bhuird, Ben Macdhui, Cairngorm and Ben Bainac rise all around it, and their rugged bases skirt its edges, except at the narrow outlet of the Aven. Its water is luminous and of great depth. At its western end is the famous Clachdhian or Shelter Stone, an immense block of granite, forming the broad shoulder of Ben Macdhui. The stone rests on two other blocks imbedded in a mass of vegetation, and forms a cave sufficient to contain twelve or fifteen men. Here, where the queen of the storm sits, at a distance of a score of miles from all human abode, the summer visitor to the wild and sombre glories of Loch Aven takes up his abode for the night. Between Inchrory and the Gairn is Loch Builg, upwards of a mile long and about half that breadth, at an altitude of 1566 feet. In these mountainous regions are a number of smaller lochs, even the names of which are unfamiliar to most people, for in these wastes of rugged hill and moor they are seldom seen by human eye.

6. Geology.

Geology is the science which investigates the history of the earth. Its object is to trace the progress of our planet from the earliest beginnings of its separate existence through its various developments to its present condition. Geology does not confine itself to changes in the inorganic world; it shows as well that the present race of plants and animals are the descendants of other and very different races which once peopled the earth, and that, however imperfectly characteristic types of the animal and vegetable kingdoms have been preserved or may be deciphered, materials exist for a history of life upon the earth. While receiving help from other sciences, geology claims as its peculiar territory the rocky frame-work of the globe. In these materials

lie the main data of geological history, arranged in chrono-
logical sections, the oldest lying at the bottom and the
newest at the top. Geological structure has of course intimate connection with the quality and natural fertility of
soil, so that its influence on agriculture and food production
is direct and intimate.

The predominant rocks in Banffshire are granite, quartz
rock, mica-slate, clay-slate, syenitic greenstone, graywacke,
graywacke-slate, Old Red Sandstone, metamorphic limestone, and serpentine.

Gneiss, and to a greater extent mica-slate, form the
lowest stratified rocks running nearly south-west from the
coast between Cullen and Portsoy to the upper valleys of
the Fiddich, Deveron, and Aven. Generally they are fine-
grained slaty rocks, and form low rounded mountains,
decomposing into soils of considerable fertility. In many
places the mica-slate alternates or passes into quartzite,
which differs from it chiefly in the almost entire absence of
mica. Quartzite in a more independent form is seen on
the coast between Cullen and Buckie, and forms also the
Durn Hill, near Portsoy; the Bin of Cullen; the Knock
Hill, and much of the high ground to the south. Where
it prevails the soil is not fertile. Connected with this series
also beds of limestone are common and have been quarried
in many places, as in the Boyne, Fordyce, Keith, Mortlach
and Kirkmichael. The limestone extends a long distance
and may be traced, indeed, from Portsoy to Glentilt in
Braemar. In Glenlivet, limestone may be found in almost
every stream and under every field; some years ago lime
kilns were to be seen on almost every farm in that district.
Clay-slate occurs in considerable abundance; in some places

perhaps merely a finer variety of mica-slate; in others coarser in texture, the so-called greywacke. Large masses occur near Boharm, and from Dufftown south to Kirkmichael. It also forms the north coast from Knock Head by Banff, Macduff, and Gamrie to Troup Head, rising on occasion into bold lofty cliffs.

The Whale's Mouth, Cullen
(*Shewing archaean quartzites, on which rests Old Red Sandstone, with glacial deposits above.*)

Resting on these rocks, Devonian or Old Red Sandstone and conglomerate beds are seen in a few places. The Morayshire beds, for instance, cross the Spey near Fochabers, running along the coast to Buckie, and in this district, in the Tynet Burn in particular, they have yielded many characteristic fossil fishes. Gamrie Bay has been eroded out of Old Red Sandstone by the sea, the action of which has been resisted more successfully by the harder rocks of Gamrie Mhor and Troup Head, but between Troup Head and Pennan there are precipices of Old Red Sandstone 400 to 500 feet high. The Old Red Sandstone of this district has been long famous for fossil fishes, and for

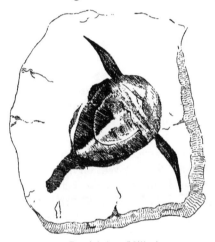

Pterichthys Milleri
(*A ganoid fossil fish of the Old Red Sandstone.*)

exposed sections of the strata which show clearly the succession of the beds.

In the southern extremity of the county, near Tomintoul, another deposit of Old Red Sandstone occurs over an area of nine by two-and-a-half miles.

Indications of more recent formations are seen in the chalk flints common in the vicinity of Portsoy, and in the oolite fossils found in the brick-clays at Blackpots, Boyndie.

The raised beach, with recent shells, more than 200 feet above the sea-level, near the old church of Gamrie, is also interesting, while Boyndie, Portsoy, Sandend, Findochty, also provide examples of recent and raised beaches. The surface of the county generally is covered with masses of boulder clay and stratified drift beds, the materials often derived from a considerable distance. Some of the granite boulders are many tons in weight.

The most important igneous rock is granite. This rock, a portion of the great central mass of the Grampians, forms the mountains in the extreme south of the county round the sources of the Aven. Ben Rinnes also is granite, and smaller masses are seen in Glenlivet and other localities. Small bosses of granite occur at Longmanhill, in Gamrie, and at Cairn of Ord, in Banff. There and in other areas the rock has been quarried to some extent. The well-known "graphic granite" forms a vein on the coast near Portsoy. It gets its name from the quartz and felspar crystals appearing on the polished surface, like rude letters. Syenite, a compound of hornblende and felspar, covers a large district, running south from near Portsoy to Rothiemay. The serpentine of Portsoy is said to have been at one time extensively wrought. It is the locally famous Portsoy marble, from which Louis XIV had chimney-pieces made for the Palace of Versailles. The district of Portsoy has been described indeed as a geological museum, showing a unique section on a small scale of the rocks which compose the Highlands of Scotland. Within a comparatively small area we find serpentine, diorite, Labradorite, graphite, quartzite, clay-slate, mica-schist, grey granite, graphic granite, and other rocks.

7. Natural History.

In recent times—recent, that is, geologically—no sea separated Britain from the Continent. The present bed of the North Sea was a low plain intersected by streams. At that period, then, the plants and the animals of our country were identical with those of Western Europe. But the Ice Age came and crushed out life in this region. In time, as the ice melted, the flora and fauna gradually returned, for the land-bridge still existed. Had it continued to exist, our plants and animals would have been the same as in Northern France and the Netherlands. But the sea drowned the land and cut off Britain from the Continent before all the species found a home here. Consequently, on the east of the North Sea, all our mammals and reptiles, for example, are found along with many which are not indigenous to Britain. In Scotland, however, we are proud to possess in the red grouse a bird not belonging to the fauna of the Continent.

The physical conditions of Banffshire, with a coastline formed by rocky cliffs, and sand and shingle beaches, with ground rising in the interior to all elevations up to over 4000 feet, drained by rivers such as the Spey, the Deveron, the Aven, and the Isla, and by innumerable small streams, and with every variety of soil and exposure, are highly favourable to plant life. Wild flowers abound everywhere, and ferns, mosses and other plants invite the attention of the botanist. The Banffshire Field Club has published a Flora of the county giving a list of over one thousand plants, and supplementary lists have been from time to time added, so that means are available for a fairly exhaustive acquaintance with the subject. Records of 23 species and varieties of ferns have been made, of 82 grasses and 28 sedges, and of 44 local mosses.

British plants have been divided into groups based on their relative distribution throughout the country. Banffshire has about five-sixths of the Scottish type and four-fifths of the Highland type; most of the British type are common, but of the distinctively English type there are few.

In some of the southern parishes with their large tracts of elevated moor, and containing some of the highest mountains in Scotland, great areas of heather make the hillsides glow in autumn in radiant colours, and provide food for large stocks of grouse. Extensive masses of whin and broom form a striking feature of the landscape in many districts. Avenside has lovely stretches of birches, and by the banks of all the streams there is in their season an abundant and brilliant show of flowering plants. The woodlands, the riversides, the coast, the hills and moors, and the roadsides all yield numerous subjects, and to those interested the county in its endless variety of native and imported vegetation provides a highly favourable field of study.

Particularly in the vicinity of mansion houses, many noble trees are to be seen. Remains of the primeval forest, which extended over a large part of the county, are to be found in some places, and huge trunks mostly of fir and oak are frequently dug up in the mosses from under deep "lairs," *i.e.* beds, of peat. Plantations consist for the most part of mixed hard wood and conifers. Ash trees, copper and green beeches, the elm, sycamores, gean, hazel, the

Valley of Aven

Scots fir, larch, silver fir and spruce each provide magnificent specimens. In the Flora of Banffshire seventeen varieties of willow are mentioned. A silver fir in Duff House grounds, blown down in the winter of 1916, had about 140 annual rings and at a foot from the ground was four feet across. In this area and elsewhere there are many fine trees, of the planting of which the second Earl of Fife in 1787 wrote in the *Annals of Agriculture*, how, within a period of thirty years, "about seven thousand acres of bleak and barren moor had been cloathed with thriving and flourishing trees...it was generally believed that no wood would thrive so near the coast; I have proved that to be a mistake; my park is fourteen miles round, and I have every kind of forest tree from thirty years old, in a most thriving state; and few places better wooded." In the glen at Cullen House there is a Pinetum which was planted about 1865 and which contains many fine specimens. In some districts on the coast, trees exposed to the furies of the north wind are often bare on their northern side and bend their branches and tops towards the opposite quarter, good examples of which will be found in the Fir Wood, Banff, and in a plantation near the viaduct at Cullen, where the trees at the northern end have their projections turned southward by the blasting winter gales from the North Sea.

In many districts the mountain ash forms a beautiful feature of the winter landscape. Early in the last century much planting was done and if the necessities of the Great War led to cutting down on a considerable scale, large forest areas are left for the natural adornment of the country side and to act as an ameliorative influence on the rigours of the northern climate.

The fauna of Banffshire is wonderful in its variety. The different kinds of birds are reckoned to be 228 in number. In the corries of the mountains in Glenaven and the south, the golden eagle still breeds, although the white-tailed eagle has gone from the other extremity of the county, on the cliffs of Troup. Large areas of moors are the homes of the red grouse; the partridge is common; in a few places pheasants are preserved; black game are found in several districts, while on the Cairngorms and Ben Rinnes, particularly the former, ptarmigan are fairly abundant. The rocks and cliffs of Gamrie are the habitat of immense numbers of sea fowl, including the kittiwake (kittie), the razor-bill auk (coulter), the guillemot (queet), and the puffin (tammy norrie). In caves along the coast small colonies of the rock dove are to be found, and the carrion and hooded crow are very common. At least four species of owl occur, and probably the most common bird of prey is the kestrel. Cormorants are frequently seen along the coast in winter and there are several small heronries in the county. In spring and autumn large flocks of wild geese pass screaming overhead in their seasonal migration; lapwings (teuchats) are abundant, and the voice of the landrail is heard all over the county where cultivation prevails, although in upland districts it is not so common. Few large birds are so numerous as the sea-gull, which seems of late years to have sought an inland home in ever increasing numbers. Vast numbers breed on the cliffs of Gamrie and Troup. The beautiful great northern diver is sometimes driven to the coast through stress of weather, and northern gales occasionally drive ashore isolated specimens of the little auk.

In the extensive forests mostly above the 1000 feet level (Glenaven of 39,000 acres, Glenfiddich of 33,000) the red deer has its home. In the corries of the hills the fox also breeds, but he is kept in check because of his raids on sheep flocks. Otters are captured occasionally by the banks of streams. Badgers are rare but are sometimes found. The brown hare abounds in most districts, heavy bags of white hares are got in various parts of the uplands, and rabbits are numerous. Roe deer are found in several areas. The larger rivers yield excellent salmon- and trout-fishing; and the brown trout is abundant in all the numerous mountain streams.

8. Along the Coast.

The seaboard of the county is pleasantly diversified. There are many picturesque rocks and in the eastern extremity towering cliffs, with, at intervals, such fine areas of sand as would be a powerful factor in the fortunes of a southern watering place.

Beginning at the Burn of Tynet, we go by Tannachy Sands to Portgordon, the most westerly village. Across the Burn of Gollachy, a low beach of sand and sea-washed stones leads by Arthur's Point to Buckpool, the western division of the burgh of Buckie, divided from Buckie proper by the Burn of Buckie. Passing on, we reach Portessie, with the Slough Hythe. Slough preserves the old and still common designation of Portessie, which has also been called Porteasy. Off the sands of Strathlene is Portessie Bay, skirted by the attractive line of the Craigenroan Rocks.

The coast now becomes increasingly imposing and picturesque in the bold headlands it presents to the furies of the northern ocean. Inland we skirt the Muir of Findochty, part of which, now planted, known as the Baads, is believed to have been the burial place of the Danes slain in battle by the victorious Scots under Indulphus.

The prosperous fishing burgh of Findochty, with its Crooked Hythe, spreads itself along the seashore and on the hill above. Onwards to Portknockie and beyond, the coast consists of high cliffs, rugged rocks, picturesque and far-stretching caves, the whole seeming to be the embodiment of sullen power and resistance. Of the many Portknockie caves, the most familiar is probably Farskane's Cave, named after the proprietor, who, in 1715, retired into it, along with two other gentlemen, to avoid trouble during the Earl of Mar's rebellion. The Bow-fiddle rock, familiar to every visitor to the coast, is in close proximity to Portknockie, and a little farther eastward the county reaches its most northerly point in the rugged heights of the Scaur Nose.

Rocks known as Toshie's Long Craigs are at the opening of the beautiful Bay of Cullen; and we pass along Cullen sands and links, by the Boar Craig, Round Craig and the Three Kings—outliers of the Old Red Sandstone—to the royal burgh of Cullen. Just at the entrance we cross the Burn of Cullen or Deskford, flowing through a fertile valley from the heights of Deskford. Beyond the harbour of Cullen, a footpath goes by way of Muckle Hythe and the Maiden Pap to Portlong, and thence to the bold promontory of Logie Head, near which is an area of sea-shore sand that has sharp sonorous qualities.

On a rocky eminence overlooking the sea, between Logie Head and Crathie Point, are the ruins of Findlater Castle. From Crathie Point the rocky coast goes on to Garron Point. Between the latter and Redhythe Point, is the Bay of Sandend, with the hamlet of Sandend to the west and adjoining an extensive area of the finest sand, through which the Burn of Fordyce enters the sea. From Redhythe Point onwards to Portsoy are many bold and picturesque rocks, the haunt of the geologist as well as the artist. East of Portsoy, at the Links Bay, into which falls the Burn of Durn, is St Columba's Well, and also the reputed site of St Columba's Chapel, both close to the shore. Inland there extends one of the most fertile districts of the county—the Boyne—of old a great forest-region:

> Fae Culbirnie t' the sea
> Ye may step from tree to tree.

After Cowhythe Head, we go by Old Hythe to Craig of Boyne and Boyne Bay, with the outlet of the Burn of Boyne. A few hundred yards from the sea are the massive ruins of Boyne House, while at the Craig of Boyne, a precipitous rock has the shattered remains of a still more ancient stronghold.

From Boyne Bay, the coast, still rugged and rocky, leads to the prosperous fishing-village of Whitehills. The coast-line now runs north till it terminates in Knock Head, where the Saut Stanes have hurled more than one goodly ship to its doom. Tradition says that the grey rat was first imported into this region from a vessel wrecked on this reef.

A sweep inland, with the sea skirting the sands and links of Banff, marks out the Bay of Boyndie. Close by the shore are the ruins of the old church of St Brandon.

Banff—from Hill of Doune

A field here is known by the name of "Arrdanes" and on the rising ground immediately to the east is "Swurd-danes," names believed to carry the remembrance of the position of two divisions of invading Northmen, armed with arrows and with swords. At the western extremity of the links the Burn of Boyndie enters the sea; opposite part of the links are a series of low jagged rocks known as the Tumblers. Banff is approached by way of the Elf Kirk, the Black Rock, the Babes, the Boot and the Broad Craig; and close by the harbour are Meavie Point and Feachie Craig. Banff Bay, beautiful for situation, washing the base of the Hill of Doune, and receiving the waters of the Deveron, lies between Meavie Point and the Collie Rocks off Macduff, and by its shores are the Rose Craig, once the residence of a retired poet of the Stuart Court; Craig Gilbert; the large and far-reaching bar of sand and stones formed by the action of the sea and the river; and the rocks at the Palmer Cove.

From this point onwards for the ten miles to the eastern extremity of the county, the coast is skirted by a ledge of stupendous rocks, in some places of sheer descent of over 500 feet to the sea and everywhere precipitous. Near Macduff is the picturesque Howe of Tarlair, with its fantastic rocks, and its "waters," in the efficacy of which Johnny Gibb of Gushetneuk and his friends had an abiding faith. Old Haven and its vicinity receive the Burn of Cullen and the Burn of Melrose, and inland from it is one of four great openings in the rocks, which here serve as outlets for the water of the interior, and which, branching off or widening as they recede from the sea, become straths or valleys. The largest of the four is the Den of Afforsk, while the other two are at Crovie and Cullykhan.

Hell's Lum and Devil's Peat Stack, Tarlair

The Bay of Gamrie is formed by the jutting into the sea of the mighty headlands, Gamrie Mhor (536 feet) on the west and Troup Head (396 feet) on the east. Between these points is the large open bay, one of the deepest in the Moray Firth, and provided with a fine natural breakwater in the Craigendargity rocks. At the shore of the bay, the steep and rugged rocks that form the headlands retire a little, leaving room for the village of Gardenstown, and no more. The collection of houses clings in a wonderful way to the side of the sea braes. At some points of the steep ascent one could almost fancy one might peep down the chimneys, and indeed, so abrupt is the rising of the ground in some parts that a house of three storeys may have them all ground floors, one entrance being at the front, another at the back, and the third at an end. Artists love to paint the picturesque and straggling village, with its miniature lanes, its narrow streets and the headland which prevents any great extension of the place. Half way up the rugged side of Gamrie Mhor are the ruins of the old church of Gamrie,

> An old, old church, the pride of the place,
> The pride of the north countree.

So sang the late Sir William D. Geddes, Principal of Aberdeen University, who, in his early days, was parochial schoolmaster of Gamrie.

> Half up the ribs of a bold giant hill
> That washes his feet in the sea,
> And looks like a king o'er the watery world,
> Lo! a patch of greenery,
> Westward and northward the crags rise high,
> To shield it from injury,
> And there, looking down on the beautiful bay,
> Is the Churchyard of Gamerie;
> Oh well do I love the sweet, sweet slope,
> Where it sleepeth solemnly.

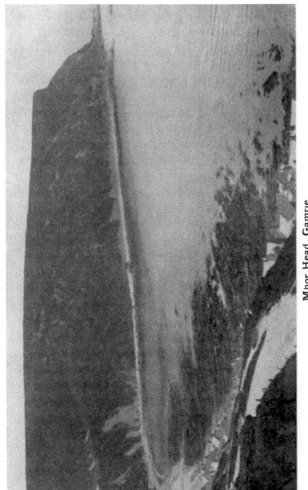

Mhor Head, Gamrie

Half a mile east of Gardenstown is the hamlet of Crovie, a single street built close to the sea, at the base of Troup Head, and, except where it faces the sea, surrounded on all sides by bold and picturesque cliffs. In the vicinity the coast-line is one of the grandest and most picturesque in Scotland. Parts of it are inaccessible to the foot of man, while others bend just enough from the perpendicular to admit a carpeting of green sward, and here and there are traversed by a winding footpath like a staircase, which few but native cragsmen are venturesome enough to scale. Several caves occur among these gigantic masses of rocks. One is 50 fathoms deep, 60 long and 40 broad, with a subterranean passage to the sea, about 80 yards long, through which the waves are driven with great violence in a northern storm, the spray escaping in what has the appearance of dense smoke—hence its name of Hell's Lum. Another, known as the Needle's Eye, is a passage, through a peninsula, of about 150 yards from sea to sea, along which a man can with difficulty creep. At the north end of this passage is a cave about 20 feet high, 30 broad and 150 long. The roof is supported by immense columns of rocks, and the effect, after one has crept through the narrow way, is exceedingly striking. The wonders of the Devil's Kitchen and the placid beauty of Cullykhan Bay lead one on to the Burn of the Tore, otherwise the Burn of Nethermill, and here in a wonderful panorama of rocky scenery the counties of Banff and Aberdeen join hands.

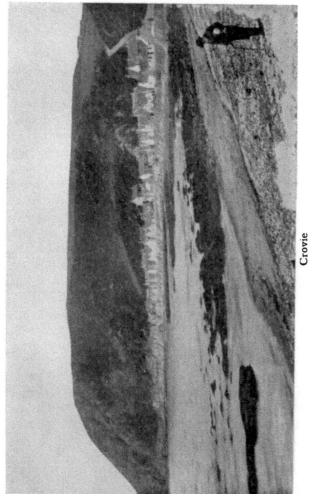

Crovie

9. Climate.

If latitude alone determined climate, Scotland would experience the rigours of Southern Greenland; but thanks to its projection into the Atlantic with the genial south-west winds, it enjoys a climate of moderate heat and cold. Banffshire, while open towards the milder west and to some extent protected from the colder east, has a high latitude, an exposure sloping to the north and facing winds from the north polar seas, and barrier heights to south and south-west. Yet its climate is marvellously favourable.

Although there are parts where winters are severe, and, lingering, too often chill the lap of spring, there are areas on the seaboard and in sheltered parts of the middle of the county, where conditions are in a way surprising when regard is had to the northern situation. At Banff, for instance, where the sunshine record is among the highest in the kingdom, the mean temperature in 1915 was 45·7 degs., in 1916 46·1 degs., and in 1917 45·8 degs. In these three years the maximum temperatures were respectively 74 degs. (May 24 and Sept. 8); 77 degs. (July 26); and 80 degs. (Aug. 6), while the minimum temperatures were 16 degs. (Dec. 5); 23 degs. (Dec. 9); and 15 degs. (Mar. 9). July and August are the warmest months and January the coldest. In 1915 the average daily sunshine period was 3 hrs. 35 mins., in 1916 it was 3 hrs. 6 mins., and in 1917 3 hrs. 48 mins.

In 35 years, 1883–1917, in which rainfall readings have been taken at Banff, the average works out to

Cullykhan Bay

28·59 inches per annum, varying from 22¼ inches in 1904 to 34½ inches in 1895. April is the driest month, followed by June and February; the wettest months are October and November. The general relation between rainfall and elevation above sea level is well shown on rainfall maps of Banffshire. A fringe along the coast has less than 30 inches a year, the line marking the fall of 35 inches curves just south of Dufftown, the higher region in the extreme south of the county has a still higher yearly average.

The most prevalent winds are south and south-east. The strength of the northerly winds is shown by the crouching of trees and shrubs in exposed situations along the coast as they bend away from the sea.

Along the coast snow does not lie long, but it is otherwise in such districts as the Cabrach, Glenrinnes, and Kirk-michael. Delightful as the summer often is there, the winter is frequently long and severe. In not a few years there is an anxious race between the maturing corn crops and the on-coming of the winter frost and snow.

10. The People—Race, Language, Population.

As the Ice Age came to an end, Neolithic men, *i.e.* men with the finely polished weapons and tools of the New Stone Age, followed the reindeer northward. Their presence in Banffshire is denoted by the existence of flint arrow-heads. These men—often designated Iberians—were small, dark-

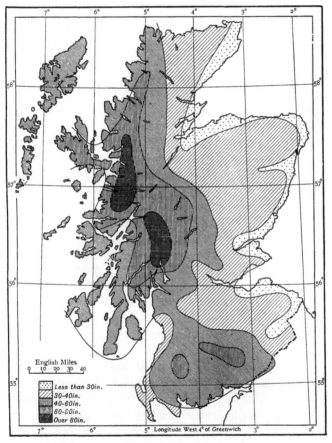

Rainfall Map of Scotland. (After Dr H. R. Mill)

haired, long-skulled. They have today their representatives outside Britain in the Basques. Names of places such as rivers and mountains are, like flints, very permanent traces of inhabitants, and the word *Isla* has been held to indicate Iberian influence. Later came the Gaelic invaders and the period of bronze tools and weapons. Then followed the age of iron, the time of the coming of the Brythons or British. In early historic times Banffshire, like the regions to west and east, formed part of the kingdom of the Northern Picts. The name *Fiddich* has been declared to be identical with the name of one of the Pictish subdivisions. The precise relationship of Picts to Iberians, to Gaels and to Britons, is problematical; but the Pictish tongue appears to have been akin to British (Welsh) not to Gaelic (Irish). *Pit* and *for* (as in *Pitlurg* and *Fordyce*) are Pictish elements; while British or Pictish are *Spey, Keith,* and *aber* as in *Aberchirder*. *Knock* is Gaelic, and so are *bal* and *inver* in names like *Ballindalloch* and *Inveraven*. English is evident in *Sandend, Lintmill,* and the first syllable of *Kirk-michael*.

The language of the English-speaking settlers has ousted the Gaelic. Or if there be still a few old people in Tomintoul who can converse a little in Gaelic, assuredly the old language, as far as concerns Banffshire, will die with them. In the Minute Book of the Synod of Moray (1715) it is recorded that the Synod considered that "Irish was necessary in the case of the minister of Glenlivet," but it does not appear that he, either at that time or afterwards, was a Gaelic-speaking minister. It is the view of Rev. R. H. Calder, the present minister of Glenlivet, that Gaelic probably became extinct in that district about the close of the eighteenth century. It held its ground more tenaciously in the neighbouring parish of Kirkmichael. There is evidence that in 1794 it was the dominant, if not the sole, language at the Tomintoul markets, and there is reason to believe that it continued to be the dominant language at these markets till the middle of the nineteenth century. The late Rev. James Grant, minister of Kirkmichael (1843–1896), was a bi-lingual preacher and a good Gaelic scholar. On the half-yearly pre-communion Fast-days, his custom was to conduct a short Gaelic service before the English service. By and by he would say to the minister who often assisted on these occasions—"Another of my Gaelic hearers gone since you were here last," and again, "Another of my Gaelic hearers gone since you were here last," and so on until they were all gone. The Gaelic service was finally discontinued about 1893.

The vernacular of Banffshire, like that of Aberdeenshire, belongs to the north-east division of Scottish dialects. Both counties have the characteristic of sounding *f* instead of *wh* in certain words; as *fah, fahn, fahr* for *who, when, where*. "Fahr are ya gain?" (Where are you going?) "Fah fuppit th' folpie?" (Who whipped the whelp?) *Folpie* shows one of the favourite forms of diminutive so common in Banffshire. Another is *ik*, and still higher, *ikie*; as *beast, beastie, beastik, beastikie*. The idea of diminution is carried still further by the adjective *wee*, by doubling the *wee*, and by adding yet again *little*, as *little wee, little wee wee*. Diminution is expressed by the word *bit*, always used in the construct state, as *a bit beastie*, and by the word *nyaff*, as *a nyaff o' a mannie, a nyaff o' a doggie*. The word *horse*, for example, may be used in descending scale from *horsie* through a multiplicity of degrees to *little wee wee horsikie*.

Within the county four divisions of dialect have been distinguished—the fishing, the lower, the middle, the higher. The fishing population accent the last syllable, throw the ictus on the last word of the sentence, and lengthen vowels. The lower district is marked off from the middle by a softer pronunciation and a slight lengthening of some vowels. Thus *bone, stone* are *been, steen* in the lower district, but *behn, stehn* in the middle. In the middle district *meal, peat, beast, beat* are sounded *mehl, peht, behst, beht,* while in the lower they are as in standard English. The upper district has been influenced in its accent by Gaelic.

The population of Banffshire at the census of 1911 was 61,402. Since the first census was taken in 1801, the highest point of population was reached in 1891, when the number was 64,190. Ten years later that was reduced by 2702, and in 1911 the total was again reduced by 86. The preliminary count in 1921 shows a decrease of 4109. Ten years ago, 46·4 per cent. of the population was enumerated in burghs and 53·6 in landward districts; and while the burghal population showed an intercensal increase of 4·3 per cent., the landward population showed a decrease of 3·7 per cent. The extra-burghal part of the county is divided into two Public Health Districts—Banff and Keith. The population of the former remained almost stationary during the intercensal period, but in the latter the population decreased by 9·4 per cent.

Of nine burghs then in the county, one only, Buckie, increased in population to an appreciable extent; in 1911 its population was 8897. The only parishes which showed a material increase of population were Rathven, the parish which contains the burghs of Buckie, Findochty, and Portknockie, and the village of Portgordon; and Boyndie, which contains the village of Whitehills.

The number of males returned as having some remunerative occupation was 18,064, and of these 32·9 per cent. were connected with agriculture and 21·5 per cent. with fishing, so that 54·4 per cent. of the male population in employment was absorbed in these two industries.

Of the total population 60,385, or 98·3 per cent., were returned as being of Scottish birth, and 75·9 per cent. of the whole had their birth-places within the county. Three hundred and seventy-eight were returned as able to speak Gaelic as well as English and none as speaking Gaelic but not English. Of the 378 only 89 were born within the county, so that the theory of the sister of the Antiquary, Grizel Oldbuck that because a man "cam' in his youth frae the braes of Glenlivet" he must needs know Erse, certainly does not hold good today.

The area of the county is 403,053 acres. In the Banff district of the county there are 125,645 acres and in the Keith district 277,408 acres, while in the former district the population in 1911 was 20,432 and in the Keith district 12,453, the population figures being exclusive of the nine burghs then in the county. In the Banff district there was on an average one person per six acres, and in the Keith district there was an average of one person per 22 acres.

The following table gives population particulars by families for the 22 civil parishes in the county and deals therefore with Banffshire as a whole:

Civil parishes	Area in acres	Separate families	Acres per family
Total ...	403,053	13,411	30·054
Aberlour ...	14,779	479	30·854
Alvah ...	12,537	245	51·171
Banff ...	6,072	1,066	5·696
Boharm ...	16,751	194	86·345
Botriphnie...	9,461	135	70·081
Boyndie ...	6,950	426	16·315
Cabrach ...	34,079	107	318·495
Cullen ...	881	528	1·669
Deskford ...	8,156	135	60·415
Fordyce ...	17,211	871	19·759
Forglen ...	6,251	133	46·992
Gamrie ...	17,047	1,459	11·684
Grange ...	15,048	295	51·0102
Inveraven ...	48,967	441	111·036
Inverkeithney	7,641	133	57·451
Keith ...	19,395	1,503	12·904
Kirkmichael	75,337	227	331·881
Marnoch ...	14,961	555	26·955
Mortlach ...	34,222	702	48·749
Ordiquhill...	4,756	136	34·8705
Rathven ...	23,182	3,404	6·8102
Rothiemay	9,369	237	39·532

11. Agriculture.

Agriculture is pursued in a most enterprising and enlightened manner, and both in arable farming and in the excellence of farm stock, the county takes a high position. Along the coast the soil consists mostly of sand and loam, the latter by far the more predominant, and these in several districts are blended with a proportion of clay soil. The arable surface along the coast lies in general upon a free open bottom, while that of the interior is mostly a light

black soil on a hard bottom, retentive of water, hence one
of the causes of the lateness of the crops in these districts.

Up to the middle of the eighteenth century improvements in agriculture were few. The arable lands were
divided into outfield and infield. To the infield, which
consisted of the acreage nearest to the farm house, the
whole manure was regularly applied; the only crops cultivated on it were oats, bere and peas, and the land was kept
in tillage as long as it would produce two or three returns
of the seed sown. When the field became so reduced and
so full of weeds as not to yield this return, it was allowed
to lie in natural pasture for a few years, after which it was
again brought under cultivation and treated in the same
manner. The outfield lands were wasted by a succession of
oats after oats so long as the crops would pay for seed and
labour. They were then allowed to remain in a state of
absolute sterility, producing little else than thistles and
other weeds till, after having been rested for some years,
they were again brought under cultivation and a few scanty
crops obtained. There are authenticated cases of fields in
Alvah and Boyndie which carried respectively 12, 14, and
19 crops of oats in succession. The system of farming
pursued was clearly described by Alexander Garden of
Troup, writing in 1686. The land as stated, was divided
into "in-field" and "outfield." The in-field was kept
"constantly under corne and bear, the husbandman dunging
it every thrie year, and if he reap the fourth corne, he is
satisfied." The outfield was allowed to grow green with
weeds and thistles, and after four or five years of repose
was twice ploughed and sown with corn. Three crops
were generally taken in succession and then, or as soon

as the soil was too exhausted to repay seed and labour, reverted to thistle and weeds. That this system was regarded as completely satisfactory, is shown by the old proverb:

> If the land be thrie year oot and thrie year in,
> 'Twill keep in good heart till the Deil gaes blin.

If credit for the change is to be given to one man more than another, it is due to James, sixth Earl of Findlater and third Earl of Seafield (d. 1770). He was an enthusiastic agriculturist, and practically transformed the face of his extensive territories, a sober eulogist writing that to him appertained "the exclusive merit of introducing into the North of Scotland those improvements in agriculture and manufactures, and all kinds of useful industry, which in the space of a few years raised his country from a state of semi-barbarism to a degree of civilisation equal to that of the most improved districts of the south." It was he who, about 1754, introduced in the north the system of alternate husbandry. He took some of his farms into his own possession, set about cultivating them in the most approved manner then known in England, under the oversight of experienced men from the south, and in a few years improved such farms as Craigherbs in the parish of Boyndie and Colleonard in the parish of Banff as well as the fields about Cullen House in a manner then unknown in these districts. He further granted to some of the most intelligent and substantial tenants leases for two 19 years and a lifetime, under which they became bound to enclose and subdivide a certain portion of the farm with stone fences or ditch and hedge during the first 19 years, and in the course of the second 19 years to enclose the remainder, while they had to sow grass seeds on a certain number of acres within the first five years of the lease. He was the first also to introduce turnip husbandry and thus pave the way for the home-feeding of cattle during winter. Other landed proprietors acted on a similarly enlightened policy. James, Earl of Fife, granted leases to improving tenants of from 25 to 30 years, with an ample allowance for building houses and dykes; and Alexander Garden of Troup followed a similar practice.

By 1812, on the larger farms where long leases had been granted, turnips were being laid down with manure, and grass seeds were being sown with bere or oats. A further improvement was effected when the broad-cast sowing of turnips gave place to drill husbandry. In course of time the eight- or ten-owsen plough was abandoned for an implement of greater tillage power hauled by horses, while with the improvement of roads the double carts (carts, that is, with a shaft and a trace horse to one cart) gave place to single-horse carts.

The system of holding land under lease for a period of years still prevails. In the uplands the arable area rises from the valley and ascends the hillside till it abuts on the heather; and in these high-lying places account has to be taken of losses through stress of weather and by game, so that in some seasons quantities of reliable seed have to be imported from more favoured districts, all circumstances that are, as a rule, reflected in the amount of the rental.

Wheat used to be grown to some extent, but in this northern climate the quality was often indifferent and the produce variable; and that, together with low prices and the poor feeding qualities of the straw, has led to its practical disappearance. Oats of the potato variety are chiefly grown; Sandwich oats, and more recently, black oats, are also cultivated, while of late years the large American varieties have been introduced. The standard weight is 42 lbs. per bushel. A large part of the oat crop is manufactured locally into oat meal, but much of it goes also to Newcastle and Leith, a considerable part for consumption in the hunting districts. The barley grown in the county is not often of the bright and attractive colour that is desired by brewers, but it finds a large and remunerative market at local malt distilleries. The turnip crop is vital to the industry of a county that is essentially a stock-rearing and feeding area. During the long northern winter, cattle kept for breeding purposes and young stock get little save turnips and oat straw, thriving magnificently upon such a diet, while cattle that are in course of finishing have supplies of feeding cake and second qualities of grain. Hay is for the most part grown only for local needs. Flax, raised in small quantities in every quarter of the county a century ago, is now seldom seen.

There are in the county 3418 agricultural holdings of an average size of 46·8 acres. Of that number 2333, or 68·26, have an area of from 1 to 50 acres; there are 1059 holdings of between 50 and 300 acres, and 26 farms of over 300 acres.

The area of the county, excluding water, is 403,053 acres; and in the year 1919 there were under corn crops and rotation grass 126,377 acres. The following figures show the cultivation of crops at different periods, and the average yearly yield over a period of ten years:

	Year 1882 acres	Year 1910 acres	Year 1919 acres	Average yield per acre over 10 years
Wheat	231	—	23	34·37 bushels
Barley	8,385	7,042	7,458	34·53 ,,
Oats	53,880	49,902	52,226	38·07 ,,
Rye	176	68	43	—
Beans	196	147	52	28·50 ,,
Peas	46	44	24	21·90 ,,
Total corn crops...	62,914	57,203	59,826	
Potatoes	2,582	1,909	1,829	6·43 tons
Turnips	25,353	21,469	20,527	17·55 ,,
Rotation grass ...	65,655	67,175	66,551	27·45 cwts.
Permanent pasture	10,122	11,275	9,198	18·04 ,,

The county is very wealthy in its pure-bred farm-stock. Of the native Aberdeen-Angus breed (indigenous to the north of Scotland and now found, by its own great beefing merits, in every agricultural country), Banffshire possesses the most famous collection in the world—that of Sir George Macpherson Grant, Bart., of Ballindalloch, one of the oldest herds of the variety and as famous on American ranches, on the estancias of the Argentine, on the veldt of South Africa and in the New Zealand bush, as it has for many years been in this country. All the world over this great collection at the confluence of the Spey and the Aven is regarded as the fountain-head of the breed. There are other Aberdeen-Angus herds of great excellence in the county. The "great intruder," the Shorthorn, has also found a favoured home. Eden and Rettie were among the earliest seats of the breed in the north of Scotland, and in the case of this as of the other great cattle breed, many of the local herds are well known in national showyards and in the foreign trade. The commercial cross cattle of the county are of a high standard, and in the most discriminating meat market of the world, that of Smithfield, they are included in the number of select cattle classed as "Prime Scots," and uniformly bring the extreme price of the day. On several occasions cattle bred and fed in the county have won the blue ribbon of the fat stock world—the champion-

Aberdeen-Angus Bull

ship of the London Smithfield Show. At the Scottish Fat Stock Show and at the show of the Smithfield Club in 1919, the champion and reserve champion of both came from Banffshire herds on Speyside, a circumstance unexampled in the agricultural annals of any county in the Kingdom.

Of late years an increased amount of attention has been devoted to Clydesdale breeding, and in the production of

strong active geldings for city use, the county enjoys a high
reputation. Sheep, too, are a valuable farming asset. There
are some select flocks of Leicesters and Cheviots, but there
are larger numbers of black-faced and cross-breds, which
spend the summer months on the hills and in the hard
weather are wintered in the pastures and surplus turnips of
the low country. The possessions of the county in farm
live-stock at different periods have been:

			1882	1910	1919
Horses	...	...	7,943	9,450	8,531
Cattle	...	...	40,957	44,227	44,923
Sheep	...	...	49,236	68,342	59,819
Pigs	...	...	3,629	3,098	2,805

The difference in the physical character of the two local
government districts in the county is well illustrated by
figures of cultivation and live-stock. In the level and fertile
Banff district there are 125,645 acres and in the more
mountainous Keith district there are 277,408 acres, while
the principal figures of crops and stock in 1919 were:

		Wheat acres	Barley acres	Oats acres	Potatoes acres	Turnips acres
Banff	...	23	6,145	29,189	1,239	12,225
Keith	...	—	1,313	23,037	590	8,302

		Rotation grass acres	Permanent grass acres	Horses No.	Cattle No.	Sheep No.
Banff	...	37,324	2,663	4,834	26,992	13,447
Keith	...	29,227	6,535	3,697	17,931	46,372

The value of the crops grown in the county in the years
mentioned is of interest. In the case of the principal field
produce the fiars prices struck were as follows, barley and

oats being both first quality and the price per imperial
quarter, while the oatmeal is per boll of 140 lbs.:

	Oats		Barley		Oatmeal	
	s.	d.	s.	d.	s.	d.
1882 ...	21	4	27	0	15	6
1910 ...	16	6	23	0	11	9
1919 ...	58	8	105	1	42	0

12. Distilling and Mining.

Banffshire takes a leading place in the United Kingdom
in the production of malt whisky.

In the Highlands of the county smuggling was in former
times very prevalent. High taxation and the interference
with cottage stills encouraged illicit distillation; and Glenlivet, celebrated for generations for the quality of its
whisky, was famous as a home of the smugglers. It has
been said indeed that smuggling houses were scattered on
every rill all over its mountain glens. In the year 1823,
when a troop of excise officers invaded Glenlivet, they
found in the glen 200 "sma' stills" in active operation.
The kegs were conveyed over the hills on horseback to Perth
and Aberdeen; and many stories are still current of events
connected with the traffic. It is on record that, when in
1824 a "legal" distillery was established in Glenlivet, such
was the opposition of the smugglers that for some time the
proprietor had to carry firearms for his protection. The
fame of Glenlivet whisky is so great that distilleries many
miles distant from the Livet attach the word "Glenlivet"
to their own distinctive name if they be situated within
the Glenlivet excise area. As a matter of fact the name of

Glenlivet has come to be, all the world over, a synonym for the liquid product of Scotland.

The 23 distilleries in Banffshire have a yearly valuation (1920–21) of £9412. They cluster mostly around Spey and Fiddich, there being seven in the parish of Mortlach, four in Aberlour and three in Inveraven, that is, fourteen in three contiguous parishes. Some of them are of old foundation. One has been at work since 1786, and quite a number of them were founded in the second decade of the nineteenth century. One of them is the largest Highland malt whisky distillery in Scotland; its yearly valuation is £1200, and in 1898 the value of its buildings, plant, etc. was put at £96,240.

Although there is no mining industry in the county in its popularly understood sense, various native products are worked to some little extent. The burning of lime is carried on at several places; and considerable quantities of the manufactured article are sent out of the county. In Kirkmichael and Glenlivet lime kilns formerly were very numerous but they are now practically all in ruins. In a recent past, limestone was quarried and burned with peats in many places, and, after slaking, the lime was spread upon the land for manure.

In the granitic areas of Upper Strathaven, crystals called Cairngorms are found in cavities in the rock and were at one time sought for among the loose debris. The Cairngorm differs from colourless quartz or rock-crystal in the presence of oxide of iron or manganese, to which it owes its colour. It is much in request as an ornamental stone. The yellow variety is not unfrequently called topaz, although quite different from the true topaz, which it resembles chiefly in colour, having neither its hardness nor its brilliancy. The topaz, however, sometimes occurs along with the Cairngorm.

Interesting minerals found in the county include magnetite, chromite, and asbestos in Fordyce; fluorite near Boharm, at Keith, and on the Aven; cyanite and chiestolite in clay-slate at Boharm. Many years ago attempts were made to work a vein of sulphuret of antimony near Keith. In the parish of Kirkmichael a vein of iron oxide (red haematite), two miles in length, occurs near the Lecht. It was worked at first for iron ore from 1736 to 1739 by the York Buildings Company. The ore was conveyed on horseback across the Aven at the Ford of Carnagaval, near Tomintoul, to Abernethy, and smelted with wood at Culnakyle. In 1841 a pit was sunk 85 feet deep to ascertain whether it would be worth while extracting manganese ore, with which the iron ore was mixed. It was found in sufficient quantity and quality, and machinery was erected to crush the ore. It was conveyed to Garmouth, and thence to Newcastle to be prepared for use in bleaching, and for a time it brought £8 per ton; but after the discovery of the utility of chlorinated lime as a bleaching agent the price of manganese ore fell to £3, and the mine had to be abandoned. At Arndilly, on the shoulder of Ben Aigan, iron ore has been worked on a small and chiefly experimental scale. Near Tomintoul there is a bed of slate stone and a slate quarry has been wrought here close by the banks of the Aven, producing good gray slates and pavement slabs.

The report for 1919 of the Chief Inspector of Mines states that in Banffshire 53 quarries were at work, employing 66 persons, so that the industry is of restricted importance. Of igneous rock there were quarried that year 23,314 tons, of granite 1520 tons, and of limestone 10,956 tons.

A large mass of diluvial clay forms the upper part of the Knock Head to the east of Whitehills, in Boyndie, and has long been wrought for the manufacture of bricks and tiles. At Tochineal, in the parish of Cullen, a bed of fine lias clay has also been wrought for many years.

13. Fishing and Fishermen.

It is highly creditable to the enterprise of the fishermen that the line- and the herring-fishing resources of the county are the largest in Scotland. The greatest herring fishery in the world is that prosecuted in Scotland, and in its contribution to that fishery, Banffshire takes the leading place.

Up to 1907 all fishing vessels belonging to Banffshire had a distinctive number which was prefaced by the letters BF. In that year the seaboard was divided into two districts of registration, Banff and Buckie, the fishing craft in the Banff Fishery district retaining the BF designation, while newly acquired boats in the Buckie Fishery district began a fresh register under the letters BCK.

The hey-day of the sail fishing fleet occurred perhaps in the eighties and in the earlier years of the nineties of last century. The Banffshire fleet of that day had a supremacy in numbers and in efficiency that was unchallenged in Scotland. The large white letters "BF" on the broad expanse of the brown sails of the yacht-like zulus were to

be seen on every sea around Britain, and their presence was
a vital influence in the fortunes of a fishing that was of
national importance. Seventy years ago the scaffie, un-
decked for the most part, was the boat in common use in
Banffshire. That type gave way to the Fifie boats, first

Zulu Boat

built in Fifeshire, while both were displaced by the zulu
boat, which for long dominated the position. This type,
matchless for sea-worthiness and sailing power, became very
popular, and for years a first-class zulu boat was the best
equipped and most highly valued vessel in the northern
fishing fleet.

The increasing size of these craft and the growing productive power of the trawling industry led to a development in Banffshire that still holds good. Lines were practically abandoned, and the whole year was devoted to the capture of herrings in all the seas round the British Isles.

The influential part taken in activities of the kind by craft from the county finds fitting illustration in official figures. At the English fishing of 1913, for instance, there took part 1163 Scottish boats, and of these 454 belonged to the county of Banff, while of the total value (£763,256) of fish landed by the Scottish fleet working that season in England, the share of Banffshire boats was £311,384. In the same year, of 159 Scottish boats that worked on the Irish coast, 88 belonged to Banffshire and of the total yield (£40,572) of Scottish boats, the contribution of Banffshire crews was £21,690.

In the early years of the present century another great development was the introduction of steam-power to the fishing craft, and now the county has a larger and more valuable fleet of steam drifters than any other county in Scotland.

The story of progress in the present century reads indeed like a fairy tale. Thus in 1900 the Banff Fishery district had no steam vessel; the Buckie district had three, the value of which, with their gear, was put at £8094. Fourteen years later the number of steam drifters in the county had increased to 398 and their value was put at over one million sterling. At the same time sailing boats decreased in value and in number: in 1900 there were in the Banff district 463 and in the Buckie district 731, figures that nine years later fell to 372 and 443. Since that time the

Steam-drifters, Buckie

same tendency has increased; and if the prophecy of a
Portknockie fisherman may not prove true that the sail
boat here may become as much a thing of the past as
Caesar's galleys, the trend of the industry is certainly all
in that direction. In the meantime the process has been
held somewhat in check by the installation in a number of
the zulu boats of motor engines and by the provision of
other new craft similarly fitted. These vessels, as occasion
demands, may use either the wind or the motor. In the
year 1913, the year preceding the outbreak of war,
when matters were normal, fishing resources in the county
were:

Propelled by sail or oar.

		No.	Value with gear
Banff	...	288	£ 49,234
Buckie	...	410	155,142

Motor Boats.

Banff	...	18	5,378
Buckie	...	12	15,147

Steam Drifters.

Banff	...	93	215,785
Buckie	...	276	766,480

That gave a total of 369 steam drifters owned in the county.
In the same year there were in all Scotland 876 vessels of
the type, the county of Aberdeen, with its three Fishery
districts of Aberdeen, Peterhead and Fraserburgh, coming
next to Banffshire with 272, or fewer than the number
owned in the single district of Buckie. Thus in 1913 the
fishing plant of Banffshire consisted of:

		No. of vessels	Tonnage	Value with gear
Banff	...	399	7,507	£270,397
Buckie	...	698	18,753	936,769
		1,097	26,260	£1,207,166

The value that year of fishing plant and gear on the East Coast of Scotland—on the West Coast the figures are negligible—was £4,736,508. Deduct from that the valuation of the fishing plant of Aberdeen, consisting for the most part of trawlers, and we reach a value for the East Coast of £3,506,294. Of that total Banffshire claimed more than one-third: so that its supremacy in potential herring-fishing capacity is unchallenged in Scotland, and we must remember that by far the greater part of these wealthy resources accrued to the county within the previous 14 years, and had been derived entirely from the democratic herring. There is the curious accompanying fact that the herring fishing in the county has declined of late years. No large fleet of boats makes any of the towns its seasonal headquarters, and the landing of herrings at county ports is confined to some extent to week-ends when boats return to replenish their fishing gear or for a general outfit.

The development of the fisheries led to large extension of the boat-building yards in the county and that industry has in consequence reached an important position. In the increase of population it has been reflected most definitely in the parish of Rathven, the headquarters of the fishermen of Banffshire. In 1861 its population was 8240; and in 1911 it was 15,995, so that in the course of fifty years the population was doubled.

In several ways the fisher folk are a race apart. Their difference of dialect has already (p. 48) been noticed. Remarkable, too, is the way their communities have maintained a continuity of family names. In a recent roll of voters, confined to male householders, all fishermen, there were in the *quoad sacra* parish of Gardenstown 17 Nicols,

At the Lines, Whitehills

19 Wisemans, 26 Wests, and 68 Watts. Macduff with a few Patersons and Watts, had 17 MacKays and 20 Wests, while in Banff the Woods numbered 27. Two names predominated in Whitehills—Lovie and Watson, numbering respectively 18 and 19.

Portsoy had 8 Mairs, 11 Piries, and 14 Woods, while in the hamlet of Sandend no fewer than 26 of the heads of fishermen's families belonged to the great family of Smith. Cullen had 33 of the name of Gardiner and 55 of the name of Findlay, while in Portknockie there were 20 Piries, 24 Slaters, 47 Woods, and 84 Mairs, or 175 heads of families of four names. Findochty went even one better, for it had 182 fishermen householders with four names between them—Campbell 24, Smith 35, Sutherland 39, and Flett 84. In Buckie, east of the Burn, there were 15 Coulls, 29 Jappys, 69 Murrays, 116 Smiths, and 128 Cowies, while in the western division there were 23 Coulls, 28 Geddeses, and 47 Reids. In Portgordon there were 21 Coulls, and 32 Reids. The circumstance explains, of course, the habit that is almost universal of having a tee name as well, a name by which the fisherman is more familiarly known than that in which he has been registered. The multiplicity in a small community of instances of the same name makes it not only a convenience but a necessity. How else, among scores of Cowies, of Coulls, of Jappys, Woods and Slaters, many of them of the same Christian name, is one to differentiate? So accustomed are many of them to be recognised only by what may be called their acquired names, that they are addressed, not by their Christian names, but as Johndie, Saners, Pum, Gyke, Dottie, Smacker, Dumpy, Bosan, Cockie, Bo, and so on.

The classic instance that illustrates the point has been told by Mr Joseph Robertson. A stranger had occasion to call on a fisherman in one of the Banffshire coast villages of the name of Alexander White. Meeting a girl, he asked, "Could you tell me far Sanny Fite lives?" "Filk Sanny

At the Cod Nets, Whitehills

Fite?" "Muckle Sanny Fite." "Filk muckle Sanny Fite?" "Muckle lang Sanny Fite." "Filk muckle lang Sanny Fite?" "Muckle lang gleyed Sanny Fite," shouted the stranger. "Oh! it's 'Goup-the-Lift' [= Sky-starer] ye're seekin'," cried the girl, "an' fat for dinna ye speer for the man by his richt name at ance?"

14. Shipping and Trade.

One of the most significant changes in comparatively recent years in the commercial arrangements of the county is the vanishing of commercial ships, especially sailing ships. Some sixty years ago the shipping industry of Banffshire employed much capital and many men on sea and on land: today it may well be said not to exist.

Some of the larger ships owned in the county seldom saw their port of registration. They were employed largely on foreign service—breaking records on a voyage from China with the new crop of tea, or in the emigrant trade (for it was no uncommon thing in those days for a ship to leave a Banffshire port direct for Melbourne or Quebec with cargo and passengers), or in a sealing expedition in the northern seas. But the majority of the ships were engaged in purely local trade—carrying coals from the Tyne, or shipping grain, cattle, salmon and products of the kind for southern ports, or cured herrings for the Baltic. In the winter months many of the ships lay up in their home harbours, a period used by the younger and more ambitious members of crews to go back to school and study navigation in order to secure the certificate of the Board of Trade. Stout sturdy fellows they were, bringing with them to the school all the romance of the far seas. Although that is now unknown, the spirit of sea adventure continues to find an outlet, and in the engine-room today of many passenger liners or cargo ships Banff lads hold responsible posts.

A Macduff gentleman, recently dead, who went to sea first in 1843, has recited the names of over one hundred

Macduff vessels he had known during his own life-time, and he lived to see a day when the town had not a schooner belonging to it. In the town of Banff, so lately as 1865, a list was made up of sixty-five persons in the town having shares in local ships, a wonderful record for a community of the size it was. There were foreign-going ships owned and registered in the town; there was a fleet of coasting ships that engaged also in the Baltic and Mediterranean trades; there were packets that sailed regularly to Leith and London; there were emigrant ships that left Banff harbour direct for Canada and Australia; there were Banff ships that engaged in the whale-fishing industry in the frozen north; and Banff ships, too, sailed west and north seal-hunting. A Banff ship in 1718 took the kirk bell of Banff, which had become "old and riven," to Holland to be recast, and when a local ship went to Bordeaux, the Town Council of the day saw to it that part of the return cargo was wine, euphemistically described as being for "the town's use." So intimate were the business relations with some continental ports that in 1730 a firm of merchants in Dantzig sent a fine brass drum, with the town's arms upon it, as a present to the burgh; and seven years later, when harbour repairs were under way in Banff, a Bordeaux merchant sent a hogshead of "strong claret" as a gift to the scheme, the said wine being rouped for eight guineas, while contributions in money were sent as well by merchants in Rotterdam. In 1781 when the Anne of Banff was captured by a Dunkirk privateer, Lord Fife had to lament " In it were all my clothes, and the whole provisions for my family in the country, with many other things." The shipping of the town and county was such that a battery with cannon was mounted above the harbour, and numbers of the inhabitants were taught to work the guns —this against depredations by French privateers—and in 1790 a Custom House was established in the burgh, the only one then between Aberdeen and Inverness.

On the absolute disappearance from Banff of its sailing commercial craft, the Banffshire Steam Shipping Coy. was formed, with its head-quarters in the town. Its first steamer, the Rosecraig, was lost on the Bell Rock; its second and last ship, the Boyne Castle, fell a victim to a German submarine while on a peaceful voyage to Newcastle for coals; the Company itself is now dissolved, and Banff is in the position of neighbouring ports in possessing no longer a mercantile marine in any form.

Portsoy had in those days a fleet of ships, and the commercial needs of the adjoining district as far inland as Huntly were served from its harbour. Cullen, Buckie, and Portgordon, particularly the last, were extensively interested in shipping. Portgordon was the recognised seaport for Keith and a large inland area, and from that little town came a race of seamen that for enterprise and skill proved inferior to none in the county. In 1841 the registered tonnage belonging to the village was 3231; between 1860 and 1870 it is believed that the majority of the male population of the village were sailors, and at one time in these years there belonged to it close on one hundred seamen who had passed the Board of Trade examination and were able to command vessels to all parts of the world, a record surely for a place of its size.

Interests so large had the natural accompaniment of shipbuilding yards in practically every port, but all acknowledged the superiority of Speymouth in the matter of ship construction, and Garmouth and Kingston yards furnished many fine vessels that carried the BF register into every sea. Probably the whole industry was in its heyday about 1857, and the mercantile navy list of that year prepared by the authority of the Board of Trade credited to the Port of Banff 142 ships, from the Corriemulzie of 606 tons, the Holyrood of 552 tons, and the Lochnagar of 379 tons to trading smacks of 15 and 22 tons.

The introduction of steam power, the coming of railways, the concentration of the herring industry at distinct centres, developments in local commerce, as well as the general speeding up of trade methods—all these had an influence in the disappearance within a comparatively few years of the sailing craft of the county.

15. History of the County.

The history of Banffshire touches national events at a number of interesting points. Whether the county was ever invaded by the legions of Rome is a matter that has been hotly disputed by antiquarians and historians, but in any case it was certainly unconquered by them. A few centuries later it formed, with what is the modern county of Aberdeen, one of the seven provinces of Pictland.

Interesting memorials still remain of the Celtic missionaries who introduced Christianity among the northern Picts. Brandon Fair, a feeing market in Banff, and "the Brannan Howe," as well as the ruins of St Brandon's Church, remind us of the famous saint. Mortlach was named after St Murthlac; and, of old, Aberlour bore the name of its patron saint, St Drostan. Alvah still possesses St Columba's well. Forglen parish used to be called Teunan or St Eunan, *i.e.* St Adamnan, the biographer of St Columba. The parish of Marnoch commemorates its patron saint in Marnan Fair. St Maelrubha was one of the most notable of Fathers of the Faith in Northern Scotland, where more than twenty places show traces of his presence. He was the patron saint of Keith, and his name is buried in Summer-eve Fair, formerly one of the most important September markets in the North. "Summer-eve" is an easy popular etymology of one of the corrupted forms of St Maelrubha's name.

While civilising influences were thus affecting the "barbarians" of the North, other movements that had been for long in progress, came to affect profoundly the national life. The pagan Vikings made descents on the growing wealth of the monastic communities, and Banffshire was the scene of three events of more or less national importance. On the muir of Findochty, in 961, the followers of Eric of the Bloody Axe and the Scots under King Indulphus, met in what is known as the Battle of the Baads. The invaders were routed, but the Scots King was slain, a collection of stones commonly called the "King's Cairn" near the farm of Woodside traditionally marking his grave. To the same epoch belongs the battle of the "Bleedy" Pits in Gamrie, where the Scots defeated the Danes with great slaughter on the top of Gamrie Mhor. The date assigned is 1004, the year inscribed on the ruins of the old church. The Scottish chief vowed to St John to build a church on the spot where the invaders were encamped if the Saint

Old Church of Gamrie and Gardenstown

would lend his assistance in dislodging them. One who wrote in 1832 recalled that three of the Danish chiefs were discovered amongst the slain, "and I have seen their skulls grinning horrid and hollow, in the wall where they had been fixed, inside the Church, directly east of the pulpit, and where they have remained in their prison house 800 years." Principal Sir W. D. Geddes has written how

> Over brine, over faem,
> Thorough flood, thorough flame,
> The ravenous hordes of the Norsemen came
> To ravage our fatherland...
> The war, I ween, had a speedy close,
> And the "Bloody Pits" to this day can tell
> How the ravens were glutted with gore,
> And the Church was garnished with trophies fell,
> "Jesu, Maria, shield us well"
> Three grim skulls of three Norse Kings
> Grinning a grin of despair,
> Each looking out from his stony cell—
> They stared with a stony stare.

To the same period is attributed the Battle of Mortlach, in which, opposite the present parish church, then a chapel dedicated to St Murthlac, Malcolm in 1010 obtained a complete victory over the Danes. It was to this chapel that Malcolm added three spear-lengths in fulfilment of a vow.

If the Reformation brought no leader from the county, and if that movement was acquiesced in rather than warmly embraced, Banffshire was intimately involved in the troubles that arose therefrom. Altochoylachan, a small stream near the eastern boundary of Glenlivet, has acquired distinction because of the battle fought on its right bank on 3rd Oct. 1594, the last struggle in the North between Protestantism and Roman Catholicism, when 10,000 Protestants under the Earl of Argyll were routed by the Catholic insurgents under the Earl of Huntly. It was a barren victory, however, for the "Popish earls" were unable to follow it up, and the King, going himself with an armed force to Strathbogie, consented to the looting of Huntly's great castle and then to its destruction. The only notable man who fell on Huntly's side was Sir Patrick Gordon of Auchindoun; Argyll lost MacLean of Mull, MacNeill of Barra, two of his Campbell cousins, and 500 rank and file. When MacLean was mortally wounded, and felt himself dying, he said to his followers in Gaelic, "Let me be buried in the churchyard of Downan, where the Saxon tongue will never be heard spoken over my grave." But MacLean, if a brave warrior, was a short-sighted prophet, for less than three centuries after, not two persons, natives of the district, could have been found who could converse in Gaelic at his grave. It was at the battle of Glenlivet that the Highland harp made its last appearance on the field of battle, brought thither by Argyll. The harp was finally discontinued in the Scottish Highlands about 1734, leaving the bagpipe the instrument of Scottish martial music.

In the time of the Covenanting troubles when the Marquis of Montrose was carrying devastation among the Covenanters of the North, we read that "from Findlater, he marched to the Boyne, plunders the country, and burns the bigging pitifully, and spoilzied the minister's goods, gear and books. The laird himself keeps the Craig of Boyne, wherein he was safe; but his haill lands, for the most part, were thus burnt up and destroyit Thereafter he marches to Banff, plunders the samen pitifully, no merchant's goods nor gear left; they saw no man on the street but was stripped naked to the skin."

The Rebellion of 1745–46 cannot be said to have met with much general sympathy in the district. From contemporary ecclesiastical and other records it rather appears, indeed, to have been regarded in the light of a nuisance, and in at least one parish the church records state, under date April 23rd, 1746, that a thanksgiving service was held "for the glorious victory over the Rebels 16th inst. where numbers of the rebel army were slain and a complete victory obtained." Sir John Cope, in 1745, on his return from Inverness, passed through Banff, having under him 2100 foot. The Duke of Cumberland left Aberdeen with the last division of his army on 8th April, 1746. Part of the Royal troops were at that time in Strathbogie and part in Oldmeldrum, and these joined him at Portsoy. He arrived on 10th April at Banff, where Lord Braco gave £250 of drink-money to the common soldiers "merely that he might with more freedom ask protection for the Houses, Cattle, Horses, and other effects of any of his friends and relations who had the misfortune of being engaged." The Earl of Findlater, at Cullen House, which had been pillaged by the Jacobites while the Earl was in attendance on Cumberland at Aberdeen, made on the same occasion "handsome provision for the troops." One incident of the Army's visit to Banff was the hanging of two men on the ground that they were spies, but the first victim is described by a later historian as "a poor innocent man"; and of both it is said that "such as knew them affirmed they had scarce wit enough to do their own country business far less play the spy." In their passage through Banff, the soldiers also "destroyed the fine episcopal chapel, cutting down the roof, burning the seats, books, pulpit and altar, and breaking the organ in pieces." Other places of worship in the county suffered a similar fate. The Roman Catholic chapel at Shenval, in the remote wilds of the Cabrach was burned, but the greatest loss to the ancient faith was the destruction of the little college of Scalan in the even more remote Braes of Glenlivet.

The name of Scalan lingers fondly in the minds of Roman Catholics in Scotland. For the greater part of the eighteenth century it was the centre of Catholic activity and over one hundred missionaries were educated within its walls. In 1713 the idea started of a seminary which would not only prepare boys for the colleges abroad, but also fully educate them for the priesthood. The place chosen was on the island of Loch Morar, but the civil disturbances of 1715 occasioned its dissolution. The work was resumed in 1717 at Scalan, a most isolated spot in Glenlivet. In May 1746 it was laid in ashes by Cumberland's troops, but although parties of soldiers were stationed in Glenlivet for nine years more, the educational work was continued and men who were to rise high in the Church received their training here. In 1789 the seminary was transferred to Aquhorthies in Aberdeenshire, and in 1829 to Blairs in Kincardineshire. The chapel of Tombae, whose priest had joined Prince Charles as chaplain to the Glenlivet and Strathaven contingent under Gordon of Glenbucket, had its contents committed to the flames. In the other ancient stronghold of Roman Catholicism in the county, the district of the Enzie, due in large measure to the protecting influence of the Gordons, the steps taken were much more lenient.

But if the Jacobite rising found small popular sympathy in the greater part of the county, it is the name of a Banff-

shire tenant farmer that stands in the front rank of those
who took part in the attempt to restore the Stuart dynasty—
that of John Gordon of Glenbucket, whose descendants,
still known by the name of "Glen," although the family
property of Glenbucket passed out of their family so long
ago as 1737, continue to have their home in an upland part
of Banffshire. In the same year, Glenbucket left Scotland
to visit the Chevalier at Rome, and papers that are quoted
lead one historian to say that "the sequence of events here
narrated makes it plain that...it was Gordon of Glenbucket
whose initiative in 1737 originated the Jacobite revival
which eventually brought Prince Charles to Scotland."
A legend lingered long in the North that George II sometimes would waken from his sleep in terror lest "de greet
Glenbogget vas goming." And this although, according
to an unknown author, believed to be a contemporary, "he
was so old and infirm that he could not mount his horse,
but behoved to be lifted into his saddle, notwithstanding of
which the old spirit still remained in him."

In a list of persons concerned in the rebellion trans-
mitted by the Supervisor of Excise at Banff, nine names of
persons from Banff appeared, one of them "now prisoner
at Carlisle"; three from Down; seventeen from Keith;
twelve from Portsoy; seven from Cullen, etc. But the
much larger number came from the district of Strathaven
and Glenlivet, where the influence of Glenbucket was
a powerful factor. About 26 landed proprietors of the
district joined in the rebellion, of whom more than one
half were Gordons.

The Forty-five brought in its wake new ideals, new
ambitions and an altogether fresh outlook to the North of

Scotland. Ancient influences in constituted society were obliterated or modified, agriculture was developed, industry grew, means of communication multiplied and from those days may be dated the modern activities of the county of Banff.

16. Antiquities.

The prehistoric period of man's existence is divided by archaeologists into the Stone Age, the Bronze Age, and the Iron Age, according to the materials of which implements of industry or weapons of war were constructed. It must not, however, be supposed that bronze implements, when first fashioned, immediately displaced stone implements or that weapons of iron at once superseded those in previous use. The different periods overlapped, and the introduction of the newer and better implements was gradual.

Of the Old Stone Age no examples have as yet been unearthed in Scotland; but of the Neolithic or New Stone Age examples are everywhere abundant. In Banffshire finely formed arrow-heads, celts and scrapers have been discovered in various districts, as Tarlair, Portsoy, the Cabrach, and Cullen. Near the Bin Hill there is believed to have been a flint factory.

The Bronze Age is regarded as having begun about 1300 B.C. Examples of bronze celts, spear heads, rings and armlets have been found in the Cabrach, at Inverkeithney, and at Auchenbadie in the parish of Alvah. The Auchenbadie armlet is particularly fine, with bold ornamentation. A most interesting relic is a boar's head of thin sheets of bronze hammered or moulded by pressure, which was unearthed in 1817 at Leitchestown in Deskford parish. It is now in Banff Museum.

Of stone circles known to have existed in the county in former days, some have disappeared, as in Rathven parish, while others are fragmentary, as at Feithhill. At Rothiemay is a recumbent stone with cup-markings.

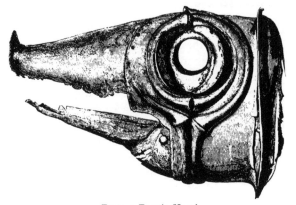

Bronze Boar's Head

Many cairns too have been obliterated in various parts of the county. Some, as at Aberlour, have, when opened, been found to contain sepulchral pottery and signs of cremation. These have been attributed to the Stone Age. Cists with skeletal remains and urns have been discovered in the Cabrach, for example, in the region of the Baads (inland from Findochty), and at East Lyne in Kirkmichael parish.

Sculptured stones occur. One is in the churchyard of Inveraven, another at Arndilly House, and a third at

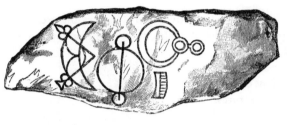

Sculptured Stone,
Inveraven Churchyard

Sculptured Stone at Mortlach

Mortlach church. The Inveraven stone has the crescent and sceptre, the triplet of circles, and the mirror and comb.

The Mortlach stone bears a cross and representations of various living creatures.

A Roman sword from what were formerly the bogs of Rettie, and Roman coins from Deskford and Dufftown have been associated with a probable march of Roman legionaries.

Cave dwellers lived at Tarlair, hut circles have been recognised in the Cabrach, the kitchen midden at the Craig of Boyne (whatever be its earliest date) has relics

Stone Pietà from Banff

coming down to mediaeval times—one has been definitely dated as of the fourteenth century.

Besides such ecclesiastical antiquities as the sacrament houses of Deskford and Cullen, a highly interesting relic (dug up in the old churchyard) of the pre-Reformation church of Banff exists in the only known example of a stone Pietà found in Scotland. It represents the Virgin in a sitting posture holding the body of Christ on her lap.

17. Architecture—(a) Ecclesiastical.

Banffshire cannot claim to possess any ancient abbey or cathedral—not even such stately ruins as Elgin. Recent days have seen churches erected of no mean architectural dignity; and the county still contains relics of interesting old church foundations.

In very early times St Moluag of Lismore founded an ecclesiastical community at Mortlach. Its history is obscure till the eleventh century. Then we hear of a battle fought in 1010 between the Scots under Malcolm and the Danes. In gratitude for his victory Malcolm added three spear-lengths to the church. Later in the century, the monastic foundation was made a bishopric by Malcolm Canmore, and the abbot became bishop. The first recorded bishop was Beathan, who was succeeded by Donert, Cormac, and Nectan. In David I's reign the see was transferred to Aberdeen. The town and monastery of Mortlach with five churches and the dependent monastery of Cloveth (Cabrach) formed part of the revenues of the bishopric of Aberdeen. The old building at Mortlach is now the parish church and its walls, though some nine centuries old as is believed, are still strong and safe. The building has during the last hundred years been enlarged and improved. The interior is of a singularly fine appearance, with interesting carved slabs and various monuments.

The ruins of the old church of Gamrie represent one of the oldest buildings in the county. Its tutelar saint was St John the Evangelist, and the date of its erection is given as 1004. It ceased to be used in 1830.

The cruciform parish church of Cullen has had a long history. The main part of the existing building was erected in 1543, according to a charter in Cullen House, which changed the previous chapel into a collegiate church for

Alexander Ogilvy's Tomb in Cullen Church

a provost, six prebends and two singing boys. A splendid example of an aumbry, with an appropriate inscription from the Vulgate, graces the north-east wall and alongside is a magnificent monument to the memory of Alexander Ogilvy of Findlater, the founder of the Collegiate Church; the monument has been described as one of the finest Gothic designs in the North of Scotland. Among other monuments is one to the memory of James, Earl of Findlater and Seafield, Chancellor of Scotland at the Union. The Seafield gallery is a piece of rare workmanship, and was erected three centuries ago when the family removed from Findlater to Cullen House.

The old church of Deskford, used for service till 1872, contains what some consider to be the finest "Sacrament House" in Scotland. It is of pre-Reformation date, an inscription bearing that it was provided by Alexander Ogilvy and Elizabeth Gordon, his spouse, in 1551. The aumbry is of carved freestone, about eight feet high. Below two angels, holding a monstrance, there is a small doorway to a chamber in the wall with a recess on either hand. At the sides and top of the doorway there is a vine branch with bunches of grapes. Across the top are the words "Os meum es et caro mea," and on the sill "Ego sum panis vivus qui de celo descendi si quis manducaverit ex hoc pane vivet in eternum. Johan sexto et cetera."

18. Architecture—(*b*) Military.

There are in Banffshire a number of interesting ruins of castles and ancient forts. The earliest strongholds were of the simplest kind—the hill or rock fort, the lake crannog, and the ha' hill. The hill fort was perhaps the earliest of all. One stands on Conval top. Sometimes the top of a rock on the coast or elsewhere was partially fortified. Of this, two rocks about a mile to the east of Tarlair, near Old Haven, are interesting examples and are among the oldest and rudest specimens of fortifications in the county.

As skill in the art of building was acquired, such structures were replaced, on the same or more suitable sites, by others more formidable, like that which hangs in ruined grandeur on the sea-crags of Findlater. In the two centuries preceding the death of Alexander III, when the settlement of Churchmen and of Norman and Saxon nobles was encouraged, architecture worthy of the name was cultivated with zeal and success, and to this period may belong the Royal fortresses of Banff and Cullen and the baronial castles of Boyne and Findlater. The styles of architecture commonly employed were the Norman or Romanesque of the twelfth, and the Early English or First Pointed of the thirteenth century. The fifteenth century witnessed the rise of a new kind of stronghold, commonly known as the Scottish Baronial Tower. These massive towers, rising floor above floor to a considerable height, and having their one door placed for safety in the second storey, afforded shelter and protection, if little else. The castles of Deskford and Inchdrewer belong to this class or are modifications of it.

The ruins of Findlater Castle, in the parish of Fordyce, are among the most picturesque in the county. A miniature Gibraltar in the day of its strength, the old castle stands on a peninsulated rock by the sea shore and still affords evidence of its former importance. John Leslie, Bishop of Ross, in his *History* (1578) describes it as "a castle so fortified by the nature of its situation as to seem impregnable," while Gordon of Straloch calls it in 1662 "deserta arx."

A tradition current locally explains why it ceased to be a family residence. While the nurse of the infant son of the lord of Findlater was walking on the sea battlement, or standing at an open window, on a genial summer day, singing and dandling the child, he, of a sudden, sprang from her arms in his glee, and disappeared in the gulf below, not, however, without a wild and vain attempt on the part of

Findlater Castle

the nurse to save him. She, too, rushed headlong into the water and perished. The baron, overcome with grief, left the castle never to return, and Findlater became, what Straloch calls it, a deserted stronghold.

On a rocky peninsula, jutting into the sea on the west side of the burn of the Boyne, some remains of buried foundations and a few masses of shapeless masonry mark the site of the original stronghold of the Craig of Boyne.

Of it nothing is now known beyond what may be gathered from a survey of its ruins. A mile from the mouth of the stream, occupying the level summit of a precipitous bank

Boyne Castle

forming the eastern side of a ravine through which the stream flows, are the picturesque ruins of a more recent Castle of the Boyne, built probably in the sixteenth or the seventeenth century, which still suggest to the beholder a strength and magnificence unequalled by any other strong-hold in the district.

The ruins of the Castle of Balvenie are close to Duff-town. The castle, of unknown antiquity, formed part of the

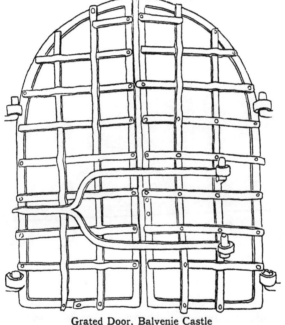

Grated Door, Balvenie Castle

extensive domains of the Comyns. On the forfeiture of the Comyns, it passed to the Douglas family. They suffered forfeiture in 1455; and the King bestowed the Barony of Balvenie on Sir John Stewart, who was created Earl of Atholl. The Atholl motto may still be read on this old castle:

"Furth Fortuine & fill thi Fettris."

Balvenie next fell to the Gordons, then to the Inneses, and then to the Earls of Fife. The ruin is in excellent preservation. The four walls, still standing, are of great strength, and are in some portions three or four feet thick. A characteristic feature is the grated iron door.

Auchindoun Castle

The ruins of the Castle of Auchindoun occupy a commanding situation on Fiddichside. According to tradition it was built in the eleventh century when the Danes were struggling for supremacy in the province of Moray. Since 1535 it has been in the possession of the Gordon family. In an issue of the *Quarterly Review* of 1816 Sir Walter Scott has told the legend of the burning of Auchindoun by "Willie McIntosh." In the ballad it is related how

> Licht was the mirk hour
>   At the day dawin'
> For Auchindoun was in flames
>   Ere the cock crawin'.

At Drumin, near the junction of the Livet with the Aven, stands the old castle of Drumin, where Argyll encamped previous to the Battle of Glenlivet and where several of the neighbouring clans joined his standard. Its founder is believed to have been Alexander Stewart, Earl of Buchan, "The Wolf of Badenoch." At all events, in 1490 the Castle of Drumin and lands pertaining thereto were disposed by Sir Walter Stewart, grandson of the "Wolf of Badenoch," to Alexander, third Earl of Huntly, and the property has ever since remained in the hands of the Earl's descendants. In later years Drumin became celebrated in the sphere of music, for it was the residence of William Marshall, described by Burns as the first composer of strathspeys of the age.

The Castle of Blairfindy, in Glenlivet, now exhibits only roofless walls. It was a square tower or keep, three storeys high, and dating from the latter half of the sixteenth century. An old distich runs:

> Glenlivet it has castles three
> Drumin, Blairfindy, and Deskee.

A farm now bears the name of the third, and but for the rhyme, it would hardly be known that a castle had existed.

Of the Palace of Banff, owned by Sir George Ogilvy, nothing remains. It was destroyed by General Monro in his fateful visit to the town in 1640. King Charles in 1641 gave Sir George ten thousand merks Scots in gold "yet too little to repair his losses," and in the following year made him Lord Banff.

The Castle of Drumin

The ancient Royal Castle of Banff has been somewhat more fortunate, for part of its walls are still to be seen, almost six and a half feet thick, and portions of its moat or fosse. It was the last stronghold occupied by English troops north of the Grampians after the battle of Bannockburn. Among those who have stayed in it are Edward I, who was on three occasions at least within its walls, and who held a Court at Banff in 1303; David II with the Queen and his sisters; Queen Margaret, wife of James III; James IV; Mary of Guise, and Queen Mary. During part of the seventeenth century it was tenanted by William Sharp, Sheriff Clerk of Banff, whose son James, the murdered Archbishop, was born there in 1618. Bought later by the Sharps, it passed in succession to the Leslies of Kininvie and the Earls of Findlater, and it remained in the possession of the latter family till 1878. The present modern house of Banff Castle was built by James, sixth Earl of Findlater and third Earl of Seafield.

The Baronial Tower or Castle of Inchdrewer, in the parish of Banff, has been referred to the reign of James IV. Inchdrewer was destroyed by General Monro, even its iron gate being sold to a countryman for five merks "whilk an hunddred pounds had not put up."

19. Architecture—(c) Domestic.

Some fine specimens of architecture are to be found in the mansion houses of Banffshire, and their richly wooded parks provide scenes of physical beauty not always associated with the north of Scotland.

Duff House is one of the most stately edifices in the lower part of the county. It is situated in beautifully wooded grounds of 165 acres, in which are some lordly specimens of copper beech, ash, elm, and other trees. Its erection was begun in 1730 by William Duff, Lord Braco and first Earl Fife. It was built after a design by William

Duff House

Adam, father of the celebrated architects of that name, at a cost of about £70,000, and is regarded as one of the happiest creations of the elder Adam. The style is purely Roman. The building has much exquisite carved work, which when the spectator is near enough is seen to be in a high degree rich, graceful, and majestic. In 1906 the late Duke of Fife and the Princess Royal made the magnificent gift of Duff House and 140 acres of the grounds to the burghs of Banff and Macduff. Since 1913 it has been held on lease by a company as a private hospital for the treatment of disorders of nutrition.

In the richly wooded policies retained in the possession of the Fife family, stands the Gothic mausoleum, now closed. It was erected by the second Earl in 1790. In the

Cullen House

chapel are old monuments with effigies, and memorial slabs with inscriptions to members of the Fife family. Outside the building is an arch over an altar with recumbent figure of a knight in armour, and on a stone beneath the arch is inscribed "This Mausoleum is erected on the place where stood a chapel dedicated to the Blessed Virgin by King Robert Bruce, MCCCXXIV. The adjacent grounds were also devoted by his Royal Charter for the building and support of a monastery of the holy brethren of Mount Carmel."

Cullen House, the principal residence of the Seafield family, founded in 1600, stands on a high rock, around the base of which circles the Burn of Cullen. Its natural situa-

Drummuir Castle

tion is striking and picturesque, and no effort has been spared in enhancing the natural beauty around. It is a castellated building of the Scottish Baronial type. The exterior has rich decorations of carving and mottoes, while some of the windows are magnificently adorned.

A beautiful feature of the upper valley of the Isla is the Castle of Drummuir erected in the late forties of last century by Admiral Archibald Duff. It is an elegant and imposing edifice set in lovely surroundings. The style is Tudor-Gothic; and its large proportions and compact form, and its castellated and embrasured roof, with the "banner tower" rising high above, give it the look of massive strength.

Above the centre of the porch looking to the east and west, there are two armorial shields, with the motto "Kind heart be true, and you shall never rue."

Forglen House is one of the most charmingly situated mansions on the lower Deveron. It was founded in 1839 and took the place of a structure erected in 1444. It is a large castellated edifice, in the Elizabethan style. Over the entrance are placed the arms of Scotland. Below these, are the arms of the owner at the time that part of the house was originally erected, the year 1577. Above the Royal Arms there is written:

Hoip of Revaird cavses gvid Service.

Under the arms of the family is the inscription—

Do veil and dovpt nocht
Althoch thov be spyit;
He is lytil gvid vorth
That is nocht envyit;
Tak thov no tent
Qvht everie man tels;
Gyve thov vald leive ondemit
Gang qvhair na man dvels.

Below this is the line—

God gyves and has nocht ye les.

The House of Kininvie is a fine old structure by the banks of the Fiddich, between Dufftown and Craigellachie. Not far from Kininvie is the House of Buchromb, which was in his later years the home of that fine old Indian soldier Sir Peter Stark Lumsden, of Penjdeh fame.

Arndilly House nestles among beautifully wooded lawns, on the east side of the Spey, on the slope of Ben Aigan, which forms an imposing background. Adorned with Grecian colonnade in front and other touches of architectural taste, it is one of the prettiest and pleasantest sylvan retreats on the Spey.

Ballindalloch Castle

Aberlour House is situated in charming woodlands near the Spey. It is of the Grecian-Doric style of architecture. It is of two storeys, and square in form, its massiveness being relieved by two well-proportioned wings. The most imposing feature is a magnificent porte-cochere that graces the centre of the building and protects the grand entrance to the mansion, supported on four elegant and massive fluted columns. There are extensive grounds all around. It is now the northern home of Sir John R. Findlay, proprietor of the *Scotsman*.

Ballindalloch Castle stands on the bank of the Aven, on a beautiful haugh, among wide-spreading trees. It is considered one of the most perfect specimens extant of the

Rothiemay House

old Scottish Castle. It bears the date 1546. The motto of the Ballindalloch family appears over the gateway at the romantic Bridge of Aven, "Touch not the cat bot a glove." The Castle grounds suffered severely from the deluge of 1829, when the mansion house itself was flooded on the ground floor to the depth of several feet. Ballindalloch Castle is the stately home of Sir George Macpherson Grant, Bart.

Rothiemay House, on the left bank of the Deveron, is surrounded by a beautiful park with many magnificent trees. The walls of the older part, eight feet thick, have been little altered from their original form, and constitute one of the most interesting series of ancient apartments in the north of Scotland.

Park House

The mansion house of Park in the parish of Ordiquhill is an elegant and handsome building, to which considerable additions were made in 1829. It is the property of the Duffs of Drummuir, who represent also the Gordons of Park.

The mansion house of Glassaugh in the parish of Fordyce occupies a finely wooded site about a mile from the sea. It is a substantial and handsome square building of three storeys of Grecian type, with a fine porch at the entrance supported by pillars.

20. Communications—Roads and Railways.

In a short perambulation of some of the main roads, we may conveniently start at the east "neuk" of the county, in the parish of Gamrie. Close by the picturesque village of Pennan, Banffshire receives from Aberdeenshire a road that, somewhat in switchback fashion, traverses the coast of Gamrie, and near Macduff joins the more important turnpike that comes from Central Buchan. At the east end of Banff Bridge the road bifurcates, going in one direction by Turriff, Fyvie, and Oldmeldrum, to Aberdeen, and in the other across the bridge. At the parish church of Banff it gives off another branch, which goes southward by Aberchirder and Bridge of Marnoch to Huntly and Donside, while the main road continues westward through Banff.

At Ordens, in the parish of Boyndie, the road again divides. One road follows the coast by Portsoy, Cullen, Inchgower (for Buckie) and Fochabers to Elgin, Forres and the north. This road all the way to Moray is probably one of the most level in the county, the only serious gradients being as it passes through Cullen and onward to the Baads. The other line of roadway from Ordens goes through Ordiquhill and Grange, skirting a shoulder of the Knock Hill, to Keith, thence, steadily ascending, to Dufftown. At the Square, while a line branches off to the Cabrach, the main roadway climbs through Glenrinnes to Glenlivet and thence through Kirkmichael, either by the picturesque, birch-lined Avenside road or by the lonely, storm-driven moor of Faemusach to Tomintoul, where it goes in one direction by a more or less precipitous way to Grantown-on-Spey by the Bridge of Brown, and in another, by one of General Wade's roads, over the mountainous Lecht Hills to the upper reaches of the valleys of the Dee and Don.

General Wade's Bridge, Glenlivet

A main line of road extends also from Portsoy to Huntly, leaving the county at Avochie; another, starting from near Cullen, traverses Deskford and Grange and joins the Banff-Keith turnpike about a mile from the latter town; while a third extends from Portgordon to Keith, all these three running in a more or less direct line from the sea southward.

The railway companies serving Banffshire are the Great North of Scotland and the Highland. The latter traverses

The Well of the Lecht

a small part of the county from Keith by Mulben into Moray. This Company had also a branch line from Keith by Enzie and Buckie to Portessie, but the rails were lifted to satisfy war needs and so far they have not been replaced.

The Great North of Scotland Company manages the other railway lines. In 1858 the railway reached Keith from Huntly; and in June 1863 the first train was run from Dufftown up Speyside, by way of Craigellachie. In the early sixties the line from Grange to Banff and Portsoy was opened. This line was extended in 1886 by Cullen and Buckie to Elgin. In 1860 the railway from Turriff reached Gellymill, near Banff, and twelve years later was carried across the face of the Hill of Doune to Macduff.

These communications by rail are with the south and the west. To the east, railway communication is not satisfactory, with the result that the distance between Macduff and its nearest large neighbour to the east, Fraserburgh, 25 miles by road, is, by rail, about 80 miles, in the form of a triangle by Inveramsay, Dyce and Maud.

Many of the outlying places are now expeditiously served by motors.

21. Administration.

In the twelfth century Scotland was divided into sheriff-doms, and the sheriffs were the servants of the Crown. Of Banffshire sheriffs, one of the earliest recorded was Richard de Strathewan. The Comyns were the hereditary sheriffs of the county previous to their overthrow by Bruce in 1308. For five years short of two centuries the Castle of Banff and the hereditary office of Sheriff belonged to the Earls of Buchan; and, when Buchan disposed of the Castle to Archbishop Sharp's family, he resigned the heritable sheriff-ship in favour of Baird of Auchmedden, in whose family it continued till 1681. Then it was conveyed to the Earl of Findlater. After the Forty-five heritable jurisdictions were abolished, and justice was then administered by sheriffs and sheriff-substitutes, nominated by the Crown. The counties of Aberdeen, Banff and Kincardine form one sheriffdom, while there is a resident sheriff-substitute in Banff.

At the head of the official county lists stand the Lord-Lieutenant, the Vice-Lieutenant, and Deputy-Lieutenants. Banffshire returns one member to the House of Commons.

The county has the two royal burghs of Banff and Cullen and the nine police burghs of Macduff, Portsoy, Buckie, Keith, Dufftown, Aberchirder, Aberlour, Portknockie and Findochty. Each of these has its own local administration save in the matter of police, in which the county is a single unit. The administration of county affairs is in the hands of the County Council, appointed by thirty-three electoral divisions, and divided into two Local Government Districts, the Lower District with twenty members and the Upper District with thirteen members. To these are added repre-sentatives of parish councils and royal burghs.

Besides the twenty-two civil parishes mentioned on page 50, there are eleven *quoad sacra*—Buckie, Enzie, Gardenstown, Glenlivet, Glenrinnes, Macduff, Newmill, Ord, Portsoy, Seafield, Tomintoul.

Banff Burgh Schools

By the Education Act of 1872, School Boards in every parish had the charge of education; but the Education Act of 1918 has now established an Education Authority, of thirty members, for the whole county to control both primary and secondary schools.

22. Roll of Honour.

Of distinguished Banffshire families whose names are connected with national as well as local history, we may mention the Duffs, the Seafield family, the Sinclairs of Findlater, the Gordons, the Abercrombys, the Grants of Ballindalloch. The noble family of Duff has an ancient and influential connection with Banffshire and the adjoining counties. The ducal line is descended from Adam Duff of Clunybeg, in Mortlach, who died in 1674, and in course of a few generations they became owners of extensive territories in the north of Scotland. In 1879 when the late Duke of Fife succeeded as Earl, the estates exceeded a quarter of a million of acres. Ten years later he married H.R.H. Princess Louise of Wales, now Princess Royal. The Duke died in 1912. Next year his elder daughter, Princess Alexandra, Duchess of Fife, married Prince Arthur of Connaught and has one son, Prince Alistair Arthur of Connaught, Earl of Macduff.

The Seafield family represents the Grants of Strathspey as well as the Findlater branch of the Ogilvys. It was a Seafield who in 1707, as Chancellor of Scotland, affixed his signature to the Act of Union with England and is said to have exclaimed, "Now, there's ane end of ane auld sang."

The fifth Earl was known as the largest planter of trees in Great Britain in the last century. By 1847 nearly 32 million young trees, Scots fir, larch and hard wood, had been planted under his direction over an area of 8223 acres. That work was continued by his successors. In 1877, in the time of the seventh Earl, in the Duthil district alone 14 million fir trees had been planted since 1866. The last Earl died of wounds in France, 1915. His only daughter became Countess of Seafield in her own right.

Banffshire was one of the districts where Roman Catholicism remained powerful after the Reformation. The north-west corner, the Enzie, has been extraordinarily fruitful in vocations to the priesthood and, small though its area be, most remarkable for the number of bishops it has supplied to Catholic Scotland. Lewis Innes, born at Walkerdale, Enzie, in 1651, became, in 1682, Principal of the Scots College at Paris. That post he resigned in 1713 to act as confidential secretary to James III, the Old Pretender. James Gordon (1664–1746) of Glasterim, Enzie, Vicar-Apostolic of Scotland from 1718, was in 1731 appointed to the Vicariate of Lowland Scotland. James Grant was born at Wester Bogs, Enzie, about 1709. While serving in the Island of Barra he was taken prisoner in 1746 as a Jacobite; but no accusation having been lodged against him and the Protestant minister of Barra and others having borne testimony to his peaceful behaviour during the insurrection, he was liberated in May 1747. In 1755 he was consecrated Bishop of Sinita and in 1766 he became Vicar-Apostolic of the Lowland District of Scotland. He died at Aberdeen in 1778. Another son of the Enzie, recognised at home and abroad as a brilliant and versatile genius, was Dr Alexander Geddes, born in 1737. Educated for the priesthood at Scalan in Glenlivet, and at the Scots College in Paris, he became priest at Auchenhalrig in Western Banffshire, where he showed such a breadth of sympathy with the Protestants that he was deposed. Aberdeen University, however, made him LL.D., the first Roman Catholic since the Reformation to receive such an honour. Between 1792 and 1800 he published a translation of the Bible into English for the use of Roman Catholics, together with Critical Remarks thereon, which exposed him to the charge of infidelity. He died in London in 1802. His poetical writings include "Oh, send my Lewie Gordon hame"—"It needs not a Jacobite prejudice," said Burns, "to be affected with this song"—and those most amusing lines "The wee Wifukie." He translated Horace's *Satires*, calling the volume *The Roman Soul transfused into a British Body*. John Geddes, his cousin, born in 1735, was President of the Scots College in Madrid, and in 1780 became Bishop of Morocco. One of his works is *A Life of St Margaret Queen of Scotland*.

Abbé Paul Macpherson was born at Scalan. He was educated at the seminary there, at Rome, and at the Scots College at Valladolid. After service in the Cabrach, Aberdeen, and Edinburgh, he was appointed in 1793 Agent of the Scottish Mission in Rome.

When Pope Pius VI was carried off a prisoner by order of the French Directory, Abbé Macpherson was employed by the British Government to secure his release, but the plan being disclosed, the Abbé was arrested, plundered and cast into prison, and the Pope died the next year at Valence, in the interior of France. About the same time, 1798, he was mainly instrumental in securing the most valuable of the Stuart Papers for the Prince of Wales, afterwards George IV. After his liberation he again served at missions in Scotland. He died in 1846, in the 91st year of his age, and the 68th of his priesthood. Rev. John Ogilvie (1580–1615) is probably the only martyr belonging to the county. He was son of Walter Ogilvie of Drum, Keith. He became a member of the Society of Jesus, and about 1614 returned to Scotland from abroad as a Roman Catholic propagandist. He was arrested in 1615 and tried and executed in Glasgow. The last of the Roman ecclesiastics we shall mention is Bishop Kyle, born at Banff in 1788. He was appointed a professor in the seminary at Aquhorthies at the age of twenty and for eighteen years he was Director of Studies. When for the first time after the Reformation Scotland was divided into bishoprics, Dr Kyle was selected to be Bishop of the Northern District and for long had his headquarters at Preshome, in the Enzie. He died in 1869, with the reputation of having been *the* builder up of his church in his diocese.

Belonging to a different communion and associated with national, not local, affairs was Archbishop Sharp, son of the sheriff-clerk of the county. He was born in Banff Castle in 1618 and murdered on Magus Muir, near St Andrews, May 3rd, 1679.

The county has naturally a goodly array of names connected with education. John Chalmers, born in 1712, son of the minister of Marnoch, was Principal of King's College, Old Aberdeen, for over half a century. Probably the most exciting event of his life occurred in 1745, when he was taken prisoner by the Jacobite army. He succeeded in making his escape after a month's captivity. Thomas Ruddiman, born at Raggal, Boyndie, in 1674, set off to compete

Archbishop Sharp

for a bursary at King's College when he was sixteen. On the way he was stripped and robbed by gypsies, his loss including a guinea given him by his sister Agnes from her small earnings; and he arrived in Aberdeen friendless and almost naked. He won the first bursary. He became schoolmaster at Laurencekirk and later obtained a position in the Advocates' Library, Edinburgh. In 1714 he published his *Rudiments of the Latin Tongue*, which at once superseded all other grammars in Scottish schools. He started business as a printer along with his brother Walter, founded the *Caledonian Mercury* newspaper, and in 1730 was appointed chief librarian in the Advocates' Library. He died in 1757. Two of the notable works he edited were the splendid edition of George Buchanan and the valuable edition of Gavin Douglas's *Æneid*. Another schoolmaster who turned printer was Dr George Chapman, born in Alvah in 1723. After acting as schoolmaster in various places, he was rector of Banff Academy from 1786 to 1792. Then, till his death in 1806, he carried on business as a printer in Edinburgh. Five brothers, all born on the farm of Ternemny, Rothiemay, exercised immense influence on Scottish education in the last half of the nineteenth century. They were Dr George Ogilvie, head of George Watson's College, Edinburgh; Dr Alexander Ogilvie, head of Robert Gordon's College, Aberdeen; Dr Robert Ogilvie, H.M. Chief Inspector of Schools in Scotland; William Ogilvie, rector of Morrison's Academy, Crieff (who died young); and Dr Joseph Ogilvie, head of the Church of Scotland Training College, Aberdeen. In the seventies and early eighties of last century, one of the best-known names to boys preparing for Aberdeen University was that of Dr William Dey, a native of Kirkmichael, rector of the Old Aberdeen Grammar School. His singleness of purpose and his upright character impressed themselves on many hundreds of pupils.

Three noted medical men were born within a year of one another. Sir John Forbes (1787–1861) was born at Cuttlebrae, Enzie. He served for a time in the Navy. In 1840 he was appointed physician to the Prince Consort and in 1841 physician to the Queen's Household. He was a Fellow of the Royal Society and D.C.L. of Oxford, and in 1853 received the honour of Knighthood, honours due in part to the books he published on medical subjects. Dr Robert Wilson (1787–1871) belonged to Banff. He was surgeon on an East Indiaman and in the course of many travels was imprisoned by Arabs. He bequeathed to the University of Aberdeen the bulk of his fortune with his library and collection of antiquities and paintings. He was founder of the Wilson Exploration Scholarship and donor of Wilson Museum at Marischal College. Sir James Clark, M.D., K.C.B., F.R.S., Bart., was born in 1788 at Kilnhillock, near Cullen, and died in 1870. After some years in the Navy and at Rome—where he published *Medical Notes on Climate*—he settled in London. In 1834 he became Physician to the Duchess of Kent, an office involving the medical care of Princess Victoria, who, when she ascended the throne, appointed him her Physician.

Of distinguished soldiers and sailors we may mention the following. General Gordon of Auchintoul, one of the Scots of distinction whose military reputation was won with Continental armies, was eldest son of Alexander Gordon, a Senator of the College of Justice. His first campaign was with the French. In 1693 he went to Russia, where Patrick Gordon of Auchleuchries was head of the army, and in course of years he rose to the rank of Major General. He was in campaigns against the Turks, the Swedes, and the House of Austria; and on his return to Scotland in 1711, he brought standards and military trophies he had taken at different times. In 1715 he acted as Lieutenant-General under the Earl of Mar and commanded the Highland Clans at Sheriffmuir. He was attainted for treason, but an error in the Act of Attainder ("Thomas" instead of "Alexander") saved his life and fortune. He escaped to France and, declining the offer of a commission in the Spanish service, he returned to Scotland in 1727 and died in July 1752, aged 81. No monument marks his last resting place at the Kirk of Marnoch. In the forties of the nineteenth century a member of the Russian Embassy in London went to Marnoch making enquiries for the purpose of having a monument erected in memory of one who had served Russia so well, but nothing resulted.

Another warlike Gordon was Sir William Gordon of Park, Convener of Banffshire, who joined Prince Charles at Glenfinnan, took part in the march to Derby and the retreat, and was present at Culloden. He escaped to France, obtained a Commission in Lord Ogilvy's Scots regiment in the French service, and died at Douai in 1751. Major-General Alexander Dirom (1757–1839) belonged to Banff. He did excellent work in Jamaica, and in India he fought against Tippoo Sultan, a campaign of which he wrote an account. Major-General Andrew Hay was born at Mountblairy, Alvah, in 1762. He was commemorated by a monument in St Paul's, London. The inscription tells that he "fell on the 14th of April, 1814 before the fortress of Bayonne in France in the 52nd year of his age and the 34th of his services, closing a military career marked by zeal, prompt decision and signal intrepidity." His wife was a daughter of William Robinson, Banff, and in 1912 her portrait by Raeburn was sold in London for £22,200. In reporting the sale, the *Times* said, "There can be no doubt that this is one of the most beautiful and attractive portraits ever sold at auction."

Rear-Admiral James Oughton of Farskane, Cullen, who died in 1832 at the age of 71, had fought against the Americans in the War of Independence, and against the Dutch. It was during the same period that George Duff of Banff began his naval career. Between 1777 and 1780 he was in thirteen engagements and became a lieutenant at the age of sixteen. He was killed at Trafalgar, where he commanded the Mars (74 guns). His son Norwich had joined him as midshipman a month before. At the date of Trafalgar Norwich was aged thirteen years two and a half months, being as far as is known, the youngest officer, and probably the youngest person, present. He had a distinguished career at sea, rose to the rank of Admiral, and died at Bath in 1862. A monument to Captain George Duff was put up in the crypt of St Paul's Cathedral, adjoining the tomb of Nelson.

Science and learning, literature and art have all their representatives among Banffshire men. James Gordon was the son of Robert Gordon of Straloch, antiquary and geographer, and became minister of Rothiemay in 1641, where he remained till his death in 1686. In 1647 he constructed a map of Edinburgh, and in 1661, a large plan of Aberdeen, Old and New. To illustrate this map he wrote a description of the two towns. Gordon also wrote a *History of Scots Affairs*. James Ferguson (1710–1776), the famous astronomer, was born at Rothiemay. When he was a herd

laddie, about the age of ten, his genius found expression in
the cleaning of clocks, and, at night, in mapping the stars

James Ferguson

with a stretched thread and beads strung upon it. First
in Edinburgh, and then in London, he won fame as a
student of astronomy. In 1761 he received from George III

a Royal pension, and was elected F.R.S. two years later. As the inventor and improver of astronomical and other scientific apparatus, and as a striking instance of self-education, he claims a place amongst the most remarkable men of science of his country. At Milltown of Rothiemay a graceful monument to Ferguson's memory was erected in 1907. The year before Waterloo a boy was born at Gosport, the son of a Fife militiaman. The boy was Thomas Edward, who settled in Banff as a shoemaker and died there in 1886. His irrepressible and inborn passion for the pursuit of natural history led him to collect many specimens and he discovered new species. He was a Fellow of the Linnean Society and of the Royal Physical Society of Edinburgh. His biography by Smiles in 1876 awakened much sympathy in his favour and a pension of £50 a year was conferred on him. The Rev. Walter Gregor, LL.D., was born in 1827 at Forgie, Keith. He was minister of Macduff, 1859–63, and of Pitsligo, 1863–95. His book on folk-lore showed much research and his *Banffshire Dialect*, published in 1866 for the Philological Society, is a valuable contribution to the study of language. The most learned son of Banffshire was Professor John Strachan, LL.D., born at Brae, near Keith, in 1862. After a brilliant university career, first at Aberdeen, then at Cambridge, with additional study in several German universities, he became, at the age of twenty-three, professor of Greek in Owens College, Manchester. Later he undertook also work on comparative philology, and the lectureship in Celtic. He died in 1907. His lectures and his publications, in Greek, Irish, Welsh, and comparative philology, showed how deep and wide was his accurate scholarship.

Journalism is worthily represented by James Gordon Bennett, son of an Enzie farmer and born in 1795. He went to America in 1819, and after hard experiences he founded the *New York Herald*, which, when he died in 1872, was (with the possible exception of Mr Greeley's *New York Tribune*) the most influential newspaper in the United States. Here may be added the name of Alexander Elder, born in Banff in the end of the eighteenth century. Going to London, he and a lad Smith from Elgin founded the publishing house of Smith, Elder and Co. Elder died in 1876.

Alexander Craig of Rosecraig was born at Banff in 1567. He went to England on the accession of James to the English throne and when he published his *Poetical Essays* he dedicated them to the King. His qualities of courtier got him a pension—"he wrote encomiastic poems in a high strain of flattery on the King and Queen"—when he retired and settled at Rosecraig, Banff.

William Gordon Stables, author and novelist, was born at Aberchirder in 1838. He studied medicine at Marischal College, Aberdeen, made several trips to the Arctic Seas, for nine years he was a surgeon in the Navy, and he was the author of more than 150 books, most of them tales of adventure. Several sculptors have hailed from Banff. John Rhind (1828–1892) belonged to an old family long connected with the town. The tasteful Biggar Memorial Fountain in Low Street, Banff, is from his plans. He was the successful competitor for the design for the memorial statue of Dr Robert Chambers, Edinburgh, and strange to say his sons William Birnie Rhind and John Massey Rhind were awarded the second and third premiums. Other works by him are statuettes on the fountain of Holyrood Palace, statue of the Earl of Kellie at Alloa, memorials of the Duke of Atholl at Dunkeld, the Duchess of Sutherland at Golspie and the Earl of Dalhousie at Brechin. Another sculptor was Alexander Brodie, born in 1830. The statue of Queen Victoria in Aberdeen is by him, and among his works are Highland Mary, the Mitherless Lassie, and Cupid and Mask. His brother William Brodie, born at Banff in 1812, became A.R.S.A., in 1851, and Academician in 1859. Among his public statues are those of the Prince Consort at Perth, Sir David Brewster and Sir J. Y. Simpson at Edinburgh, and Dr Graham of the Mint at Glasgow.

Alexander Cassie of Banff, George Smith of Fordyce, and William Hay of Ordiquill, were generous benefactors to their places of nativity. Cassie was born in 1753 and died in 1822. He had a successful career in the West Indies and later carried on a sugar refinery in London. He left upwards of £20,000 for the poor of his native town. Smith, after realising a considerable fortune in India in the eighteenth century, died on his way home. He bequeathed most of his fortune for educational purposes in Fordyce. With his name has to be linked that of Walter Ogilvie, who in 1678 bequeathed Redhyth and other lands in Fordyce to establish bursaries at Fordyce and King's College, Aberdeen. Hay had a very successful business career in Australia, was a large landowner there, and gifted to his native parish the beautiful Hay Hall.

In Elspeth Buchan, Banffshire has its one native who has founded a religious sect—the fanatical Buchanites of the West of Scotland in the last quarter of the eighteenth century. Born at Rothmackenzie in Fordyce parish, she was the daughter of John Simpson, a wayside innkeeper; and, after a wayward girlhood, she married Robert Buchan,

Archibald Forbes

a Glasgow potter. Her views and practices in the spheres of religion and life soon caused separation from her husband.

She induced the Relief minister of Irvine to adopt her opinions, for which he was deposed by his presbytery, and in 1784 the magistrates of that town expelled Elspeth and her followers. They found a resting place near Thornhill in Dumfriesshire. Mrs Buchan gave herself out to be the woman of Revelation xii, and gained much notoriety by the bogus "miracles" she wrought. She died in 1791, and the last of her sect survived till 1848.

The name we conclude with—that of Archibald Forbes (1838–1900) the war correspondent—is in many ways one of the most remarkable. His father was minister of Boharm, and his mother was Elizabeth, daughter of Archibald Young Leslie of Kininvie. Forbes's career is well-known. In the battle-fields of France, in the Balkans—he was decorated by the Tsar for personal bravery before Plevna—in Burmah, South Africa and elsewhere, he won a distinguished position in his arduous profession. To his early experiences as a soldier and his practice of journalism, were added indomitable resolution and energy as well as a fine physique capable of a large amount of endurance, qualities which stood him in good stead in many an emergency. Witness his famous ride of 110 miles in 15 hours to report at once the victory of Ulundi in 1879. In 1884 Aberdeen University made him an honorary LL.D.

23. The Chief Towns and Villages of Banffshire.

(The figures in brackets after each name give the population in 1911, and those at the end of each section are references to pages in the text.)

Aberchirder (1048), in Marnoch parish, begun in 1746 by General Gordon of Auchintoul, was constituted a burgh in 1889. The Rose-Innes Cottage Hospital, for the parishes of Marnoch and Forglen was founded by Miss E. O. Rose-Innes of Oldtown of Netherdale. In the old churchyard, two miles from the town are the fragments of a building that had been used in succession as a Roman Catholic, Episcopal and Presbyterian parish church. The adjacent existing church was built about the commencement of last century. A historic event was enacted within its walls and around them on that wintry morning of 21st January, 1841, when the seven suspended ministers of Strathbogie met to induct Mr John Edward to the pastorate of the parish. The proceedings of that day rang throughout ecclesiastical Scotland and helped to precipitate the disruption of the National Church two years later. (pp. 107, 120.)

Aberlour (1272) is, of right, Charlestown of Aberlour, from the name of its founder early in the nineteenth century, Charles Grant of Wester Elchies. The parish, of the same name, was originally Skirdustan—the division of Dustan (Drostan), its tutelary saint. The town became a burgh in 1894. The fine Town Hall was the gift of the late Mr James Fleming, distiller and bank agent, who left also £9000 to build and endow a cottage hospital for Aberlour and £500 for the erection of a suspension foot-bridge over the Spey to connect with Knockando. Aberlour has a large orphanage founded by the late Canon Jupp, a man of abounding enterprise. (pp. 19, 59, 74, 82, 100, 107.)

Banff (3821) is a very ancient place of human habitation. Standing at the mouth of a productive river, the town had fertile land around it. Through the friendly co-operation of river and sea a bar formed a harbour of refuge well suited to the wants of early navigation, and it was a member of a northern Hanse, connection with which gave valuable trading privileges. Malcolm Canmore may have made it a royal burgh: it is mentioned as royalty in 1057. A charter of Robert II,

Aberchirder—from the East

of date 1372, is still in existence. For centuries it has been the seat of the administration of law and local government in the county. Banff suffered from both parties in the stormy Covenanting times: "no merchant's goods nor gear left" is the record after a plundering by

Cross of Banff

Montrose in 1645. Soldiers were frequently in it, sometimes quartered on it because of the non-payment of taxes, sometimes, as after the Forty-five, to quell Jacobite ambitions in the district.

A surgeon with Cumberland's army in 1746 records that then the town lived chiefly by smuggling, and 30 years later it is designated as

"perhaps the gayest little town in Scotland." Wolfe, the future con-
queror of Canada, described Banff as "a remote and solitary part of
the globe....When I am in Scotland, I look upon myself as an exile."
John Wesley visited Banff three times, and wrote of it as "one of the
neatest and most elegant towns that I have seen in Scotland." Boswell

St Peter's Church, Buckie

and Johnson spent a night in it, the latter uttering a growl because
the inn windows had no pulleys. Burns stayed a night in the town,
8th September, 1787.

Byron as a boy lived for a short time with relatives at Banff and by
one indignant person was characterised as "that little deevil, Geordie
Byron." The poet Southey, who visited Banff in 1819, with his friend

Telford the engineer, then engaged in harbour extension, wrote of it as a "clean, cheerful and active little place.'

The Town-house, built between 1796 and 1800, occupies part of the site known as the Towers, which were almost the last remains of Lord Banff's Palace. The Burgh Cross is very old and is mentioned in 1542. There are a number of fine memorial stones in the old church-yard, including a monument to the family of Archbishop Sharp. Some of the quaint old houses bear inscriptions, but little is left to mark the site of the Carmelite monastery. The Public Library and Museum was erected in 1902 on the site of the "Turrets," where Cumberland had his meal store for the army. Chalmers Hospital was built and endowed from the estate of Alexander Chalmers of Clunie, merchant and shipowner of Banff and Gardenstown, who died in 1835. There are two boat-building yards in the town and the fishing industry is of considerable importance. (pp. 5, 17, 24, 30, 34, 42, 61, 63, 65, 68, 71, 73, 77, 78, 80, 84, 88, 93, 95, 103, 106, 107, 112, 115, 116, 119, 120, 121.)

Buckie (8897), the earliest fishing station in Rathven parish, is the largest town in Banffshire and has more line fishermen than any other town in Scotland. It consists of three main divisions: Buckpool, Easter Buckie and Portessie. Buckie, Buckpool, Ianstown and Gor-donsburgh were erected a burgh in 1888 and Portessie was incorporated in it in 1903. The activities of the town centre largely round the harbour, which was opened in 1879 and which, with later additions, cost the Cluny Trustees £80,000 when it was purchased by the Town Council, under whose management it has been greatly extended at a heavy expenditure. The Buckie fishermen own a magnificent fleet of fishing craft whose influence is felt wherever they operate round the British Isles. The graceful twin spires of St Peter's Church, rising to a height of 115 feet, form a notable feature of the local architecture. (pp. 23, 24, 32, 48, 49, 61, 63, 65, 68, 72, 103, 106, 107.)

Craigellachie is the railway junction of the main line Aberdeen to Elgin and the Strathspey line to Perth *via* Boat of Garten. It is beautifully situated near where the waters of the Fiddich reach the Spey and is very popular as a summer resort. One of the most beauti-ful bridges in the county is that over the Spey. The cast-metal arch and framing, so light and airy in appearance, the embattled towers which flank the abutments, the rugged and precipitous rocks over-hanging the river and the roadway, the dark deep pool below, the finely wooded eminence in the background, and the picturesque accessories which enhance the prospect on all sides, combine to render

Craigellachie Bridge a picture of no ordinary beauty and charm. It was erected in 1815 by Mr Simpson of Shrewsbury, after a design by Telford, and withstood the floods of 1829, when the water here rose to the height of 15 feet above the ordinary level. The arch is 150 feet span, and the turrets rise to 50 feet. (pp. 19, 106.)

Crovie (282), a fishing village in the very east of the county, is built close on the sea at the base of Troup Head. It stands in a picturesque neighbourhood. (p. 40.)

Craigellachie Bridge

Cullen (1992) is a town of great antiquity. It has a charter of James II, March 1455, ratifying burghal privileges granted by Robert I. It consists of two parts, the New Town and the Seatown. The predecessor of the former, which was called the Old Town, was meanly built and about 1820 was utterly demolished in order to make room for improvements at Cullen House. In March 1645 Cullen was plundered by the Farquharsons of Braemar, by orders of Montrose, and to help it in its distress collections were ordered to be made in all churches of Scotland. The municipal buildings were erected in 1822 at the expense of the Earl of Seafield; the harbour provided by the Earl of Seafield in 1817 has on various occasions since then been

Seatown of Cullen, Viaduct and Links—from Castlehill

enlarged and improved. The Seatown is inhabited mainly by the fishing community. In George Macdonald's novel, *Malcolm*, the scene of which is laid in Cullen, it is thus described: "The Seatown of Portlossie was as irregular a gathering of small cottages as could be found on the surface of the globe. They faced every way, turned their backs and gables every way—only by the roofs could you predict their position—were divided from each other by every sort of small irregular space and passage, and looked like a National Assembly debating a Constitution." Cullen House grounds contain many magnificent forest and ornamental trees. Nearly parallel to the links runs a railway embankment 50 feet in height and connecting at its eastern extremity with a viaduct over the burn of Cullen consisting of eight arches, each of 63 feet span and upwards of 70 feet in height. (pp. 23, 30, 33, 61, 68, 72, 78, 80, 84, 86, 88, 103, 104, 106, 107, 117.)

Dufftown (1626), in the parish of Mortlach, has been a police burgh since 1863. It was founded by the Earl of Fife in 1815. It is now one of the chief centres of the distilling industry in Scotland and it has a flourishing business as well in the burning of lime. A tower in the Square is a leading architectural feature in the town. The Stephen Cottage Hospital, opened in 1890, was built and endowed by Lord Mount Stephen, the most famous native of Dufftown. For many years a great power in the commercial activities of Canada, he was one of the potent forces in the construction of that work of Imperial magnitude—the Canadian Pacific Railway. Both in this country and in the Dominion his benefactions have been princely. (pp. 6, 24, 44, 84, 91, 103, 107.)

Findochty (1776), founded in 1716, was erected a burgh in 1915. Fishing is the paramount industry. In common with similar communities on the coast it has increased much in size of late years. (pp. 26, 33, 49, 68, 74, 107.)

Fordyce (297), a very ancient place of settlement, was erected a burgh of barony by charter dated 1499 and renewed in 1592. Of one of its clergymen of the early post-Reformation period, Mr Gilbert Gairdn, it is said that he "seldom went to the pulpit without his sword, for fear of the Papists." Fordyce has for long been best known for its secondary school, with its valuable bursaries. (pp. 6, 23, 60,121.)

Gardenstown (1090) was the first herring fishing station on the Moray Firth, the industry having been established in 1812. Standing on Gamrie Bay, it is in summer a favourite of artists, who seek to transfer to their canvases the rugged beauties of the east "neuk" of Banffshire, with its twin villages—Gardenstown and Crovie—its dark

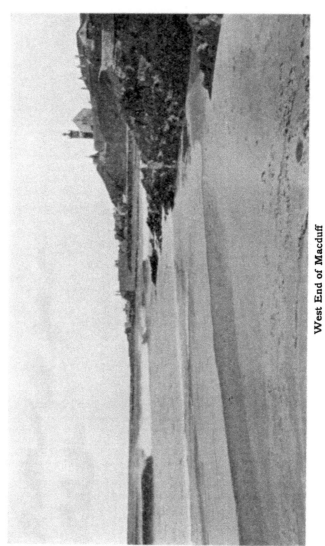

West End of Macduff

caves, its precipitous headlands, the home of a great variety of seafowl, its seething cauldrons and its gigantic chasms in the rocks through which the wind sighs and the sea waves roar. (pp. 38, 66.)

Keith (4499) is the chief agricultural centre of the county. It consists of three communities, which were united into a police burgh in 1889. Old Keith, a very ancient place, was celebrated for its annual three days' market held in September, whither traders resorted from places as far apart as Glasgow and the Orkneys. New Keith, on the Seafield property, was begun about 1750. Fife-Keith, on the other side of the Isla, was founded by the Earl of Fife in 1817. Keith is the seat of an extensive industry in meat, and has important cloth and manure factories. The town is annually the scene of one of the largest one-day agricultural shows in Scotland. The present Institute buildings date from 1889, and much good work has been done in the community by the Turner Memorial Cottage Hospital. (pp. 6, 23, 60, 72, 74, 80, 103, 104, 106, 107, 119.)

Macduff (3411) formed, of old, part of the Thanage of Glendowachie, which comprised a large area east of the Deveron. Robert the Bruce granted the Thanage and other lands to his sister's husband Hugh, Earl of Ross. Later it was for long in the possession of the Earls of Buchan. As a small fishing village, less than two centuries ago, it was known as Doune or Down—it clusters along the eastern base of the Hill of Doune—and in 1733 it was bought by William Duff of Braco, afterwards first Earl of Fife, a family that has greatly fostered the interests of the town. Through the exertions of the second Earl it was in 1783 erected into a burgh of barony. The harbour has been the mainspring of the town's activities. It was purchased in 1898 by the Town Council from the Duke of Fife and since that time a large amount of money has been spent on its extension and improvement, and it is now one of the largest, safest and most convenient harbours in the Moray Firth. An extensive addition—the Princess Royal Basin —was opened by the Princess Royal in May, 1921. The burgh possesses a fine town hall. In the vicinity is the Howe of Tarlair with its iron spa, situated amidst impressive rock scenery. (pp. 24, 36, 68, 70, 80, 103, 106, 107.)

Newmill (558) in the parish of Keith, the successor of a village of great antiquity, was founded by the Earl of Fife in 1759.

Portgordon (1369), the most westerly town in the county, was founded in 1797 by the fourth Duke of Gordon, after whom it is named. In 1874 it was provided by the Superior with a new harbour,

Tomintoul

which was enlarged and deepened in 1909. In the pre-railway era it was the port of shipment for an extensive inland area. (pp. 32, 49, 68, 72, 104.)

Portknockie (1746), founded about 1677, was made a police burgh in 1912. Its interests are almost entirely in fishing. It has a fine fleet of about fifty steam-drifters. (pp. 33, 49, 65, 68, 107.)

Portsoy (1951), erected a burgh of barony in 1550, became a police burgh in 1889. Its industries include fish-curing and the manufacture of oatmeal. (pp. 23, 25, 26, 34, 68, 72, 78, 80, 103, 104, 107.)

Sandend (316), to the west of a beautiful bay with a fine expanse of sand, is wholly devoted to fishing. (pp. 26, 34, 68.)

Tomintoul (479), standing 1160 feet above sea-level, is the highest village in the Highlands, while the distinction of the highest in Scotland falls to the Lanarkshire village of Leadhills. It was founded in 1750 on what was a bleak and barren moor, but of late years its amenities have been greatly improved; and, possessing some of the most important attributes of a Highland health resort, it has been described as "at once the breeziest, healthiest, and most primitive little town in the Kingdom." Tomintoul occupies a table-land overlooking the river Aven, with views of hills and glens and wooded passes, with the mountains of Ben Aven and Ben Macdhui in the distance. Its railway stations are Grantown-on-Spey (14 miles) by Wade's road *via* Bridge of Brown; Ballindalloch (15½ miles) *via* the Faemusach or the picturesque Avenside; and Dufftown (20 miles) by way of Glenrinnes. (pp. 6, 13, 18, 25, 46, 60, 107.)

Whitehills (1108), in Boyndie, is a thriving sea-side village, almost entirely dependent on the fishing industry. It is an important centre for line-caught white fish. A new harbour was opened in 1900. (pp. 34, 49, 61, 68.)

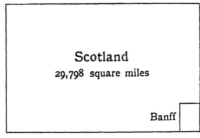

Fig. 1. Area of Banffshire (630 sq. miles) compared
with that of Scotland

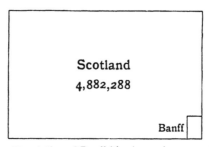

Fig. 2. Population of Banffshire (57,293) compared with
that of Scotland in 1921

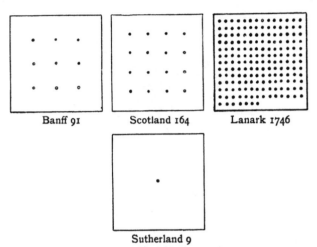

Banff 91   Scotland 164   Lanark 1746

Sutherland 9

Fig. 3. Comparative density of population to the square mile in 1921. (*Each dot represents 10 persons.*)

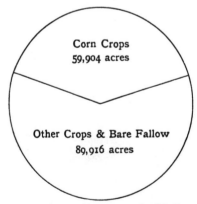

Corn Crops
59,904 acres

Other Crops & Bare Fallow
89,916 acres

Fig. 4. Area under cereals compared with that of other farmed land in Banffshire in 1919

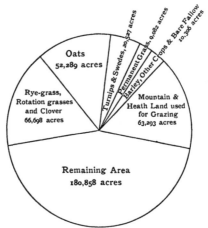

Fig. 5. Proportionate areas of land in Banffshire in 1919

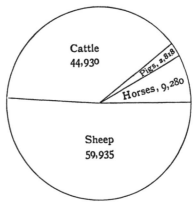

Fig. 6. Proportionate numbers of live stock in Banffshire
in 1919

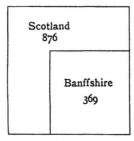

Fig. 7. Comparative numbers of steam-drifters

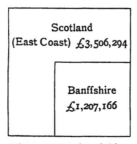

Fig. 8. Value of line- and herring-fishing craft and gear

Ingram Content Group UK Ltd.
Milton Keynes UK
UKHW010856130623
423278UK00011B/124